Disaster Response

Other Books in the Current Controversies series

Disaster Response

Debra A. Miller, Book Editor

GREENHAVEN PRESS
A part of Gale, Cengage Learning

GALE
CENGAGE Learning

Detroit • New York • San Francisco • New Haven, Conn • Waterville, Maine • London

Christine Nasso, *Publisher*
Elizabeth Des Chenes, *Managing Editor*

© 2009 Greenhaven Press, a part of Gale, Cengage Learning

Gale and Greenhaven Press are registered trademarks used herein under license.

For more information, contact:
Greenhaven Press
27500 Drake Rd.
Farmington Hills, MI 48331-3535
Or you can visit our Internet site at gale.cengage.com

For product information and technology assistance, contact us at

Gale Customer Support, 1-800-877-4253
For permission to use material from this text or product, submit all requests online at www.cengage.com/permissions

Further permissions questions can be emailed to permissionrequest@cengage.com

Articles in Greenhaven Press anthologies are often edited for length to meet page requirements. In addition, original titles of these works are changed to clearly present the main thesis and to explicitly indicate the author's opinion. Every effort is made to ensure that Greenhaven Press accurately reflects the original intent of the authors. Every effort has been made to trace the owners of copyrighted material.

Cover image © Michael Ainsworth/Dallas Morning News/Corbis.

LIBRARY OF CONGRESS CATALOGING-IN-PUBLICATION DATA

Disaster response / Debra A. Miller, book editor.
   p. cm. -- (Current controversies)
  Includes bibliographical references and index.
  ISBN-13: 978-0-7377-4134-6 (hardcover)--
  ISBN-13: 978-0-7377-4135-3 (pbk.)
  1. Disaster relief--United States. 2. Emergency management--United States.
  3. Emergency management. 4. Disaster relief. I. Miller, Debra A.
  HV555.U6D583 2009
  363.34'80973--dc22
                                            2009009039

Printed in the United States of America
1 2 3 4 5 6 7 13 12 11 10 09

Contents

Chapter 2: Has the U.S. Disaster Response Improved Since Hurricane Katrina?

Chapter 3: Should Citizens Rely on the Government to Respond to Disasters?

Yes: Citizens Should Rely on the Government to Respond to Disasters

No: Citizens Should Not Rely on the Government to Respond to Disasters

Chapter 4: How Can U.S. Disaster Response Be Improved?

Foreword

By definition, controversies are "discussions of questions in which opposing opinions clash" (Webster's Twentieth Century Dictionary Unabridged). Few would deny that controversies are a pervasive part of the human condition and exist on virtually every level of human enterprise. Controversies transpire between individuals and among groups, within nations and between nations. Controversies supply the grist necessary for progress by providing challenges and challengers to the status quo. They also create atmospheres where strife and warfare can flourish. A world without controversies would be a peaceful world; but it also would be, by and large, static and prosaic.

The Series' Purpose

The purpose of the *Current Controversies* series is to explore many of the social, political, and economic controversies dominating the national and international scenes today. Titles selected for inclusion in the series are highly focused and specific. For example, from the larger category of criminal justice, *Current Controversies* deals with specific topics such as police brutality, gun control, white collar crime, and others. The debates in *Current Controversies* also are presented in a useful, timeless fashion. Articles and book excerpts included in each title are selected if they contribute valuable, long-range ideas to the overall debate. And wherever possible, current information is enhanced with historical documents and other relevant materials. Thus, while individual titles are current in focus, every effort is made to ensure that they will not become quickly outdated. Books in the *Current Controversies* series will remain important resources for librarians, teachers, and students for many years.

In addition to keeping the titles focused and specific, great care is taken in the editorial format of each book in the series. Book introductions and chapter prefaces are offered to provide background material for readers. Chapters are organized around several key questions that are answered with diverse opinions representing all points on the political spectrum. Materials in each chapter include opinions in which authors clearly disagree as well as alternative opinions in which authors may agree on a broader issue but disagree on the possible solutions. In this way, the content of each volume in *Current Controversies* mirrors the mosaic of opinions encountered in society. Readers will quickly realize that there are many viable answers to these complex issues. By questioning each author's conclusions, students and casual readers can begin to develop the critical thinking skills so important to evaluating opinionated material.

Current Controversies is also ideal for controlled research. Each anthology in the series is composed of primary sources taken from a wide gamut of informational categories including periodicals, newspapers, books, U.S. and foreign government documents, and the publications of private and public organizations. Readers will find factual support for reports, debates, and research papers covering all areas of important issues. In addition, an annotated table of contents, an index, a book and periodical bibliography, and a list of organizations to contact are included in each book to expedite further research.

Perhaps more than ever before in history, people are confronted with diverse and contradictory information. During the Persian Gulf War, for example, the public was not only treated to minute-to-minute coverage of the war, it was also inundated with critiques of the coverage and countless analyses of the factors motivating U.S. involvement. Being able to sort through the plethora of opinions accompanying today's major issues, and to draw one's own conclusions, can be a complicated and frustrating struggle. It is the editors' hope that *Current Controversies* will help readers with this struggle.

Introduction

> *"Hurricane Katrina's main legacy . . . was widespread outrage about the poor performance of the nation's emergency management and disaster response system."*

Most Americans remember the summer of 2005 as the time when Hurricane Katrina hit the coastline of the Gulf of Mexico. Katrina was a monster storm that at times reached category five, the highest level on the scale used by the U.S. National Hurricane Center. The high winds and rain destroyed many parts of coastal Louisiana, Mississippi, and Alabama, and the storm surge—massive waves that build up under a storm—caused flooding that virtually wiped out the low-lying city of New Orleans. Katrina left in its wake almost twelve hundred dead, more than a million people homeless and displaced, and tens of billions of dollars in property and economic losses. Hurricane Katrina's main legacy, however, was widespread outrage about the poor performance of the nation's emergency management and disaster response system.

The concern about disaster response arose as millions of Americans watched television images of New Orleans citizens who were not evacuated, many of them poor African Americans, suffering without aid in the aftermath of the storm. Many people were trapped in their homes, and as the water rose, their only escape was to poke holes in their attics and climb onto their roofs, to await a rescue effort that was painfully slow and poorly coordinated. Other survivors walked many miles to reach designated shelters that soon became overcrowded, leaving people with little food or water, no electricity or air conditioning, poor sanitation, and relentless heat and humidity. It is estimated that some six hundred thousand pets either died or were abandoned because shelters would not allow animals, producing heart-breaking televised images of their plights as well. The catastrophic nature of the tragedy became even clearer in the weeks, months, and years after the storm when many people who had lost their homes still had no permanent place to live, and often no jobs, in the region they had once called their home.

Following Hurricane Katrina, the many failures in the federal, state, and local governments' efforts led to widespread criticism of the U.S. disaster response system and cries for its improvement. In particular, critics blasted the slow and incompetent performance of the Federal Emergency Management Agency (FEMA), the nation's main disaster relief agency. In addition, many were disappointed in President George W. Bush's failure to immediately take charge of relief efforts; the poor disaster preparation and management of resources by state and local officials also came under fire.

Since Hurricane Katrina, numerous changes have been made to address the problems revealed in 2005. At the federal level, FEMA was reorganized and given a more prominent role in disaster response. FEMA administrators say the agency now is better able to coordinate the activities of the various federal agencies that play a role in disaster relief. They believe FEMA is also prepared to work more effectively with both state and local authorities and aid organizations, such as the American Red Cross, in the event of a natural disaster or other emergency. Some state and local governments have reassessed and revised their disaster planning and response programs in an effort to avoid a repeat of the Katrina debacle.

Many observers believe these reforms have been shown to be effective, judging from the government's proactive response to several natural disasters that have occurred since Hurricane Katrina. When wildfires struck much of Southern California in the fall of 2007, for example, FEMA representatives were quickly dispatched to provide assistance and to work with local authorities, earning the federal government generally high marks. Also, during the massive floods in the Midwest in the summer of 2008, FEMA responded quickly and surely with sandbags, water, food, and housing assistance. Afterwards, most commentators praised the agency for its work with flood victims, helping to repair the agency's post-Katrina reputation. Similarly, most reports agreed that FEMA's response to Hurricanes Gustav and Ike late in the summer of 2008 showed improvement compared with its handling of Hurricane Katrina.

Some experts and reporters, however, question whether FEMA is truly ready for another major disaster like Hurricane Katrina or the September 11, 2001, terrorist attacks. They point out that the Midwest floods, the California wildfires, and recent storms have been relatively minor disasters in which FEMA was not really tested. In the event of another really catastrophic event, many experts wonder if FEMA would be able to establish clear communications, deliver supplies quickly enough, and evacuate and house large numbers of people for long periods of time if large swaths of residential areas were destroyed. Meanwhile, other commentators question the wisdom of even trying to undertake massive disaster relief and reconstruction efforts in regions of the country known to be the target of frequent and regular natural disasters such as hurricanes, floods, and wildfires. Perhaps, they suggest, these areas ought to be off-limits for residential housing and be used instead as nature preserves or tourist destinations, thereby reducing the high costs of repeated rebuilding efforts.

Whether ready or not, most experts agree that the United States will see more major disasters on the scale of Hurricane Katrina in the future. According to most predictions, category four and five hurricanes are expected to become both more frequent and more intense due to global climate change, which is causing a warming of the oceans that contributes to storm activity. Other weather changes triggered by global warming, such as drought in some places and increased rainfall in others, can be expected to contribute to other types of natural disasters, such as more wildfires in the western states and more flooding in wetter parts of the country. At the same time, the country faces a continuing threat of attack from international terrorists, as well as the possibility of disease pandemics like the avian flu. As a result of these threats, disaster response will no doubt remain a high-profile, controversial issue of great concern to both citizens and policy makers. The authors of the viewpoints in *Current Controversies: Disaster Response* address various aspects of this critical issue and offer ideas about how to better prepare for future disasters.

CHAPTER 1

What Are the Challenges Involved in Disaster Response?

Chapter Preface

For much of the nation's history, disaster response was a local concern, and the federal government only stepped in after the fact to help communities recover and rebuild. During the 1900s, the federal government began to enlarge its role in responding to disasters. The San Francisco earthquake of 1906, which resulted in fires that destroyed much of the city, was one of the first instances of federal government involvement. President Theodore Roosevelt pledged federal troops to help maintain order and protect government buildings, and later the federal government engaged the American Red Cross to provide aid to the region.

In 1950, Congress enacted the Civil Defense Act, the first significant piece of legislation to define the federal government's role in disasters. This law, however, was a response to the Cold War—the decades-long ideological struggle between the United States and the Soviet Union—and dealt mostly with defending the nation against a Soviet nuclear attack. Over the next couple of decades, federal responsibility for responding to natural disasters shifted among a variety of departments, agencies, and the White House. By the late 1970s, federal emergency management was handled by a complex mix of agencies sometimes working at cross-purposes.

Finally, in 1979, at the urging of state and local leaders and following the near disaster at the Three Mile Island nuclear power plant in Pennsylvania, President Jimmy Carter created the Federal Emergency Management Agency (FEMA), the first centralized agency to manage the nation's emergency planning and disaster response. FEMA helped to advance disaster response by coordinating federal disaster functions within one agency and by improving communications between the federal government, states, municipalities, the National Guard, and other entities involved in disasters. FEMA, however, was never able to completely fulfill President Carter's initial vision of an agency with the power to affect local building and zoning codes in order to prevent disasters from happening in the first place. Instead, some communities situated in disaster-prone areas, such as several located in the hurricane zones of the Gulf Coast, have repeatedly relied on federal monies to rebuild in the same locations.

FEMA was given additional responsibilities during the term of President Ronald Reagan in the 1980s, enabling the agency to give warnings and coordinate evacuations in the event of a nuclear attack. During this period, FEMA aggressively competed for national security resources and sought to expand its security role. This focus on national security issues, many people believe, caused FEMA to become less effective in dealing with natural disasters. In major disasters such as Hurricane Hugo in 1989, the agency became overwhelmed and was criticized for failing to deliver needed aid quickly and efficiently. The situation only worsened in 1992, when Hurricane Andrew hit Florida, causing billions of dollars of property damage and leaving 160,000 people homeless. FEMA's response teams lacked critical communications technology such as satellite phones, and FEMA failed to respond quickly to state officials who were clamoring for aid. President George H.W. Bush fired the director of FEMA, replacing him with Andrew Card, then secretary of transportation, to straighten out the chaotic recovery process.

After Hurricane Andrew, congressional leaders and others called for reforms. The result was the appointment of a new FEMA director, James Lee Witt, an experienced emergency manager who was a close friend of the country's new president, Bill Clinton. Over the next several months, Witt worked closely with Congress and oversaw a comprehensive reorganization of FEMA. Witt's reforms eliminated the agency's national security role and realigned it to focus exclusively on disaster mitigation and emergency management. Witt's approach has been called "all hazards, all phases"—a policy that gave priority and funding to programs and resources that could be used for various types of disasters, and which emphasized not only post-disaster response and recovery activities, but also disaster mitigation and preparation. In fact, FEMA became known during this period for its efforts to mitigate disasters; for example, it encouraged homeowners in flood-prone areas to buy insurance, reducing government recovery expenses when floods hit. Witt also placed people and equipment into regions before disasters struck and cut red tape that had often delayed aid to states and municipalities, improving FEMA's response capability. Under Witt's tenure, FEMA regained its reputation as an agency that could effectively respond to natural disasters. The number of declared disasters increased throughout President Clinton's terms of office, and people began to expect the federal government to come to the rescue in the event of a catastrophe.

However, the terrorist attacks of September 11, 2001, led President George W. Bush to change once again the focus of FEMA, directing it from natural disasters toward antiterrorism efforts. As part of the new focus, in 2002, Congress created the new Department of Homeland Security to coordinate the fight against terrorism, and FEMA was placed within the new department. During this period, many of FEMA's experienced emergency managers, including Witt, were replaced with political appointees who had very little experience with disaster response. For example, FEMA director Michael Brown's previous experience was a position with the International Arabian Horse Association. Many FEMA employees also retired, and Witt's policies slowly dissolved as the president and Congress concerned themselves with the terrorism threat.

In 2005, Hurricane Katrina tested FEMA's disaster response capabilities in much the same way as Hurricane Andrew had done in 1992, with similar results. FEMA failed to respond quickly and effectively, and communications with states and municipalities were muddled. The poor performance of FEMA in 2005 led to a new round of reforms, and FEMA has handled several smaller disasters easily, working cooperatively with local authorities. It remains to be seen whether the current federal and state disaster response system can successfully handle the next Katrina-level disaster. The viewpoints in this chapter explore some of the many challenges involved in effective disaster response.

Providing Water and Sanitation Are the Immediate Challenges Following a Disaster

Pan American Health Organization

The Pan American Health Organization (PAHO) is an international public health agency with more than a hundred years of experience in working to improve the health and living standards of the countries of North and South America.

Water, a life-sustaining element, can become the source of major concerns after a disaster. It is critical to have sufficient clean water in the immediate aftermath of an event in order to treat the ill, provide for human consumption and maintain basic hygiene, support in the work of search and rescue, and to resume normal productive and commercial activities.

Access to water is a basic human right, and implies a responsibility that goes beyond the protection of investments and is, above all, a responsibility of public health.

Increased Natural Hazards

In the current global situation, characterized by conditions of inequity and extreme poverty, environmental degradation and climate change have caused an increase in the occurrence of natural hazards such as landslides, intense rains, hurricanes, drought, fires, and earthquakes. Furthermore, rapid and un-

Pan American Health Organization, "The Challenge in Disaster Reduction for the Water and Sanitation Sector: Improving Quality of Life by Reducing Vulnerabilities," Washington, DC: Pan American Health Organization, 2006. Copyright © Pan American Health Organization (PAHO/WHO), United Nations Children's Fund (UNICEF), International Strategy for Disaster Reduction (ISDR), International Federation of Red Cross and Red Crescent Societies (IFRC), 2006. http://esa.un.org/iys/docs/san_lib_docs/DesafioDelAgua_Eng-intro.pdf. Reproduced by permission.

planned urban growth has increased the number of settlements on unstable, flood-prone, and high-risk land where phenomena such as landslides, rains, and earthquakes have devastating consequences. Socioeconomic factors increase the vulnerability of communities as well as existing infrastructure and services.

Each year more than 200 million people are affected by droughts, floods, tropical storms, earthquakes, forest fire, and other hazards. As demonstrated by recent events, natural hazards can affect anyone in any place. From the tsunami in the Indian Ocean to the earthquake in South Asia, from the devastation caused by hurricanes and cyclones in the United States, the Caribbean, and the Pacific, to the intense rains throughout Europe and Asia, hundreds of thousands of persons have lost their lives and millions their livelihoods because of disasters triggered by natural hazards. The impact of events of catastrophic magnitude on all aspects of the economy and development has been evident, in particular for developing countries.

In Latin America and the Caribbean alone, the impact of natural disasters on water and sanitation systems caused damages amounting to some 650 million dollars between 1994 and 2003.

Environmental degradation and climate change have caused an increase in the occurrence of natural hazards.

In an environment where natural hazards are present, local actions are decisive in all stages of risk management: in the work of prevention and mitigation, in rehabilitation and reconstruction, and above all in emergency response and the provision of basic services to the affected population. Commitment to systematic vulnerability reduction is crucial to ensure the resilience of communities and populations to the impact of natural hazards.

Water and Sanitation Challenges

Current challenges for the water and sanitation sector in the framework of the world's Millennium Development Goals [international efforts to reduce world poverty] require an increase in sustainable access to water and sanitation services in marginal urban areas and rural areas, where natural hazards pose the greatest risk. In settlements located on unstable and flood-prone land there is growing environmental degradation coupled with extreme conditions of poverty that increase vulnerability. The development of local capacity and risk management play vital roles in obtaining sustainability of water and sanitation systems as well as for the communities themselves.

When these factors are not taken into account, there is the danger of designing and constructing unsustainable services that progressively deteriorate and malfunction. Poor design and construction put both the community and infrastructure at risk in disaster situations.

Current challenges for the water and sanitation sector . . . require an increase in sustainable access to water and sanitation services . . . where natural hazards pose the greatest risk.

Many Different Actors in Water/Sanitation

The many actors in the water and sanitation sector (the administration, supervisors, providers, consumers, etc.) complicate the definition and assignment of functions and responsibilities. This results in confusion as to who does what regarding specific actions related to disaster prevention, preparedness, mitigation, and response. During each of these phases, each of the actions and actors have one common objective, that is, to ensure that the levels of water and sanitation service, established with local authorities and the community, can be sustained even during disaster situations.

The reduction of vulnerabilities entails multi-disciplinary work in a network with other actors in risk management, such as public ministries (in particular those responsible for public works and utilities, land planning and management, health, education, and finance), disaster management agencies, NGOs [nongovernmental organizations], the private sector, and the academic sector (universities, professional associations, research centers) fostering the development and exchange of knowledge in matters of protecting water and sanitation systems against natural hazards.

Reducing Vulnerability to Natural Disasters

On the other hand, the resistance of systems to natural disasters is an important step toward ensuring that the achievements made in increased access to water and sanitation services are strengthened in the long term, thereby realizing the goal of reducing by half, by the year 2015, the percentage of people that lack sustainable access to safe drinking water and basic sanitation. In this sense, the local activities of risk management position themselves as a tool for realizing the global challenges of providing water and sanitation services for all and at all times.

In January 2005 in Kobe, Japan, 168 governments committed to act to reduce disaster risk, and adopted a 10-year plan known as the Hyogo Framework for Action, with the objective of reducing vulnerability to natural hazards. The inclusion of criteria of vulnerability reduction to the impact of natural hazards in the water and sanitation sector is a priority activity for advancing the objectives of the global plan. Furthermore, water and sanitation systems warrant consideration as "critical" infrastructure, and as such are a priority for the efforts of disaster risk reduction, as are schools and hospitals. The loss of human life and economic and environmental losses as a result of disasters in 2005 serve to strengthen the belief that disaster reduction should be an integral part of sustainable development, and a critical factor for reaching the Millennium Goals. The water and sanitation sector must not be overlooked when addressing this challenge.

Post-disaster Temporary Housing Is Often an Unmet Need

Matt Fink, Ilyssa Plumer, Kit Radosevich, Erin Ward, and Rebekah Green

Matt Fink, Ilyssa Plumer, Kit Radosevich, and Erin Ward are students at Western Washington University in Bellingham, Washington. Rebekah Green is a research associate and grant writer at the Institute for Global and Community Resilience at the Huxley College of the Environment, Western Washington University.

Washington State is at risk for a variety of natural hazard disasters. In the last 20 years alone, the state has experienced 21 presidentially declared disasters requiring some form of disaster assistance.

Many forms of assistance were needed, and these and less traditional forms of assistance will be needed in future disasters.

One especially salient issue is the need for shelter. Major flooding, earthquakes, volcanic eruptions and other less traditional hazards may displace large numbers of people and damage residential housing stock such that immediate re-occupancy is not possible. Many displaced residents may eventually need post-disaster temporary housing.

Emergency Versus Temporary Housing

As Hurricane Katrina demonstrated, there is an important difference between emergency shelter and temporary housing.

Emergency shelter is mass temporary shelter meant to house displaced populations. It is a short-term response lasting up to 6 weeks, and addresses immediate needs. It is too expensive to maintain long-term. Some examples of emergency housing include hotels, stadiums, and churches.

Temporary housing is the transitional phase between emergency shelter and permanent housing. It may be used for several months up to a few years. It addresses more long-term needs, and allows disaster victims to re-establish their regular routines and return to a more normal, productive life.

Vulnerable Populations

Of the over six million Washington State residents that have the potential to be impacted by a disaster, an estimated 11.6% of these are below the poverty line. This makes them vulnerable to being severely impacted by a disaster.

A disaster has a more profound effect on poorer residents of a region due to financial constraints that make it harder for them to recover. These issues include not having home or renters insurance pre-disaster, the inability to secure loans, or simply not having the resources to move to another area. Furthermore, [sociologist Robert] Bolin notes that affordable post-disaster housing has been a significant factor in the slow recovery of low-income neighborhoods.

A disaster has a more profound effect on poorer residents of a region due to financial constraints that make it harder for them to recover.

Washington State also has a sizable racial and ethnic minority population, with almost a 10% percent Latino population, and a black, American Indian/Alaskan Native, and Asian population of 4.5%, 2.6% and 7.9% respectively. Research on housing-related policies and outcomes after numerous U.S. natural disasters documents consistent disparities based on race, class, and gender. [Disaster recovery experts Walter Gillis] Peacock and [Chris] Girard found that racial and ethnic minorities tend to receive insufficient insurance settlements because they are less often insured by major national carriers. [Researcher Denise] Blanchard-Boehm reported that financial constraints reduced the likelihood that African-Americans made structural improvements so their houses could withstand natural disasters, resulting in more serious damage to the homes of African-Americans. [Writer and researcher Elaine] Enarson pinpoints elderly women as more vulnerable during disasters. Similarly, temporary housing put in place after a disaster is often not designed with the needs of women and children in mind.

Existing Funding for Temporary Housing

There are two primary agencies that provide support for temporary housing following a disaster. These include the Federal Emergency Management Agency (FEMA) and the Department of Housing and Urban Development (HUD).

FEMA is charged with administering the Robert T. Stafford Disaster Relief and Emergency Assistance Act, also known as the "Stafford Act." This forms the core of the U.S. government's emergency housing disaster relief strategy. Three sections of this act are directly relevant to the development and funding of a temporary housing program.

[The Stafford Act] forms the core of the U.S. government's emergency housing disaster relief strategy.

- *Section 403: Essential Assistance* Section 403 provides funds to state and local government for meeting the immediate needs of disaster survivors by providing funds for shelter, food, water, medicine, debris removal, search and rescue, and other immediate needs. It states "The Federal share of assistance under this section shall be not less than 75 percent of the eligible cost of such assistance." This section is most relevant to supporting temporary housing needs during and immediately following a disaster.

- *Stafford Act Section 408: Federal Assistance to Individuals and Households* Section 408 of the Stafford Act is aimed directly at funding temporary housing. This grants ". . . financial assistance to individuals or households to rent alternate housing accommodations, existing rental units, manufactured housing, recreational vehicles, or other readily fabricated dwellings." This program gives aid directly to impacted households.

- *Stafford Act Section 404* Section 404 governs FEMA's Hazard Mitigation Grant Program and the Pre-Disaster Mitigation Program. These grants can be used for a wide variety of mitigation efforts and they provide up to $3 million to communities to engage in mitigation projects. Such mitigation projects can reduce the need for temporary housing through such activities as buy-out programs, strengthening programs, educational programs, and other mitigation activities.

Section 408 of the Stafford Act is aimed directly at funding temporary housing.

The Department of Housing and Urban Development (HUD) has successfully provided financial assistance to disaster stricken communities in the past. There are three different program options available; these include the direct assistance program, block grants, and mortgage programs.

- *Direct Assistance Program* The Direct Assistance Program, specifically the Section 8 Housing Vouchers and/or Project Based Rental Assistance, allows people to easily move and relocate. Households pay 30 percent of their income toward rent, while the federal government funds the remaining cost. These programs are geared towards low-income populations to help them find an 'affordable cost' level of housing. The Direct Assistance Program can assist disaster-struck populations in recovery.

Existing Programs and Limitations

In the United States, two primary strategies are typically used in post-disaster temporary housing. These strategies are 1) a rental voucher program to subsidize a displaced household in finding temporary housing within the existing real estate market, and 2) giving displaced households limited-term use of mobile homes for placement on their own lots or in temporary housing camps.

In the United States, two primary strategies are typically used in post-disaster temporary housing[:] . . . a rental voucher program . . . [and] limited-term use of mobile homes.

The rental assistance voucher program has been used extensively following such events as the 1993 Midwest floods, the [1994] Northridge [California] earthquake, and Hurricanes Andrew and Katrina.

In both the Midwest flood and Northridge earthquake, rental assistance programs worked well. With a high housing vacancy rate, displaced residents were able to be temporarily housed in rental units while being close to their original residency. This enabled them to initiate repairs and continue with pre-disaster employment, schooling and community engagement.

The rental assistance program was less successful after Hurricanes Andrew and Katrina. With low rental vacancy rates in Miami-Dade County [Florida], an estimated 100,000 displaced residents used the rental voucher program [to] find affordable temporary housing north of the County. An even larger out-migration occurred following Hurricane Katrina. With a destroyed housing stock in many Gulf Coast communities, rental units were simply not available to meet the needs of displaced residents. Instead, these displaced residents used the rental vouchers to find housing throughout the country, often far from their original homes. In New Orleans, where the problem was most acute, this out-migration resulted in the drastic reduction in the local tax base, workforce, and school enrollment. Three years after the storm, New Orleans' population had reached only 72 percent of its pre-Katrina level.

Pros and Cons of a Rental Voucher Approach:

- Rental vouchers allow displaced residents to live in a permanent structure, typically in a functioning community.

- Rental vouchers can support housing stock repair and the local economy when there is a sufficient housing stock vacancy in the region.

- Use of existing real estate markets can positively affect a recovering local economy. It also does not require purchase, shipping, or disposal of temporary housing units.

- In tight housing markets, or when the number of displaced households greatly exceeds the available vacant housing in a region, rental vouchers can encourage out-migration.

- Extensive out-migration can retard community recovery by depleting the local tax base, workforce and schools.

Mobile Homes Option

The second form of temporary housing often used after disasters is mobile home units. These units, often called FEMA trailers, are allotted to displaced households in lieu of rental assistance. These trailers have been employed following Hurricanes Andrew and Katrina, and in a more limited scope following floods and wildfires. The mobile home units, like the rental vouchers, are time limited, typically running from three to 18 months. In some cases such as following Hurricane Katrina, extensions have been granted for up to two and a half years after the initial disaster.

Mobile home trailers have been sited on individual properties, thereby allowing displaced residents to quickly return to their communities. This has also allowed displaced residents to easily access their damaged homes and either engage in or monitor repairs. In some cases where mobile homes have not fit on properties or where zoning has excluded this option, mobile home trailer parks have been set up, though often with significant outcry from surrounding neighbors.

Current FEMA mobile trailers ... can be dangerous living spaces in extreme events.

Following Hurricane Katrina, FEMA trailer camps of dozens and even hundreds of mobile homes were organized both near the devastated Gulf and in-state far from the area. Residents of these camps showed higher rates of suicide and heightened incidents of domestic violence. Later testing found dangerous levels of formaldehyde within the trailers. Displaced residents also found that FEMA trailers designed for short-term use were not sturdy or spacious enough for family occupancy over extended time periods. Furthermore, large trailer camps were often placed far from employment opportunities, shopping centers, and alternative housing options. The distance of the camps from the residents' original homes made repairs and property upkeep nearly impossible.

Current FEMA mobile trailers also can be dangerous living spaces in extreme events. Trailers can fall off their pillars or be heavily damaged from falling or floating debris in floods, hurricanes, tornados, and earthquakes.

Pros and Cons of Mobile Unit Approach:

- When placed at or near the site of the original house, use of mobile home units can speed individual home repairs and community recovery.

- Large camps of mobile units can have negative social impacts on disaster victims.

- Current mobile units pose a health risk to occupants when used for extended periods.

- Mobile home units are at greater risk of damage in hurricanes, floods, tornados and earthquakes.

- Mobile homes can be costly to store or dispose of, though they can be reused in a new disaster.

Innovations in Temporary Housing

Several temporary housing alternatives have been designed through competitions and for-profit rebuilding efforts after it has become apparent that long-term usage of FEMA travel trailers cause significant social and health problems. All have drawn on the advantages of on-site or near-site temporary housing solutions that allow residents to quickly return to their pre-disaster jobs, schools, and communities. They have also drawn from the strengths of the rental voucher assistance program by designing temporary houses that have the feel and look of more permanent housing.

The Katrina Cottage was designed as an alternative to the FEMA trailer and is intended to be placed in a home-owner's yard during the rebuilding and recovery stage. There are now many designs labeled as Katrina Cottages. They all are designed to be comparable to the FEMA trailer in cost, but with additional advantages. They are generally safer in a natural hazard event and can be used as an apartment, studio or cottage even after occupants have returned to their repaired homes. Some such designs have been marketed by major home improvement retailers. Due to current restrictions on use of FEMA funds for "permanent shelters", these structures currently are not covered under the FEMA or HUD assistance programs, though pilot programs have been implemented in some Gulf Coast communities with some community resistance.

The Katrina Cottage was designed as an alternative to the FEMA trailer and is intended to be placed in a homeowner's yard during the rebuilding and recovery stage.

Western Washington University undergraduate students have also developed a series of preliminary design concepts for temporary housing for both on-site temporary housing and temporary housing camps. . . .

Policy Needs and Future Areas of Study

Based upon previous post-disaster temporary housing experiences across the United States, it is clear that many issues remain unresolved. In order to ensure that communities in Washington State will be able to quickly and effectively recover from a major disaster—whether the result of predicted subduction zone earthquake, a major flooding event, a large volcanic eruption, or other disasters—it is important to consider potential issues in policy, funding, and implementation of a large-scale temporary housing program. Some issues for consideration include:

- *Guiding Principles.* What social, ethical, fiscal, and/or environmental principles might a Washington State temporary housing program(s) be based upon? What objectives will a temporary housing program(s) attempt to meet for both individual and community recovery?

- *Minimum Standards.* What minimum standards, and for whom, may be appropriate for temporary housing program(s) in Washington State? These may include standards for:

    - Size

    - Durability

    - Toxicity

    - Reuse

    - Labor

    - Economy

    - Resource efficiency

    - Use flexibility

    - Accessibility

- *Estimated Need.* What populations are likely to be displaced in predicted major disasters? What population demographics may a temporary housing program need to accommodate?

- *Land Use.* Do current land use zoning regulations allow for placement of temporary housing units on individual lots or in temporary housing camps? If not, what are the possibilities for post-disaster exemptions?

- *Protocol.* What state and local agencies will oversee aspects of a temporary housing program? At what level(s) of need will aspects of the program be initiated and for how long? Where and when might the use of rental vouchers, mobile units, semi-permanent small units or a combination be suitable? What mutual aid agreements, if any, should be pursued within the region?

- *Proto-Design Options.* What suite of potential proto-designs may be appropriate for Washington State, based upon guiding principles, minimum standards and estimated need? Such standards may need to consider the needs of unique sub-populations, such as:
  - Elderly households
  - Households with children
  - Disabled users
  - Households with pets
  - Cultural needs or minority populations

Natural Disasters Present the Challenge of Dealing with Emotional Trauma, Especially for Children

Philip J. Lazarus, Shane R. Jimerson, and Stephen E. Brock

Philip J. Lazarus is a nationally certified school psychologist (NCSP) from Florida International University; Shane R. Jimerson is an NCSP from the University of California, Santa Barbara; and Stephen E. Brock is an NCSP from California State University, Sacramento.

Natural disasters can be especially traumatic for children and youth. Experiencing a dangerous or violent flood, storm, wildfire, or earthquake is frightening even for adults, and the devastation to the familiar environment (i.e., home and community) can be long lasting and distressing. Often an entire community is impacted, further undermining a child's sense of security and normalcy. These factors present a variety of unique issues and coping challenges, including issues associated with specific types of natural disasters, the need to relocate when home and/or community have been destroyed, the role of the family in lessening or exacerbating the trauma, emotional reactions, and coping techniques.

Children look to the significant adults in their lives for guidance on how to manage their reactions after the immediate threat is over. Schools can help play an important role in this process by providing a stable, familiar environment. Through the support of caring adults school personnel can help children return to normal activities and routines (to the extent possible), and provide an opportunity to transform a frightening event into a learning experience.

Immediate response efforts should emphasize teaching effective coping strategies, fostering supportive relationships, and helping children understand the disaster event. Collaboration between the school crisis response team and an assortment of community, state, and federal organizations and agencies is necessary to respond to the many needs of children, families, and communities following a natural disaster. Healing in the aftermath of a natural disaster takes time; however, advanced preparation and immediate response will facilitate subsequent coping and healing.

Issues Associated with Specific Disasters

Hurricanes. Usually hurricanes are predicted days to weeks in advance, giving communities time to prepare. These predictions give families time to gather supplies and prepare. At the same time, however, these activities may generate fear and anxiety. Although communities can be made aware of potential danger, there is always uncertainty about the exact location of where the hurricane will impact. When a hurricane strikes, victims experience intense thunder, rain, lightning, and wind. Consequently, startle reactions to sounds may be acute in the months that follow. Among a few children subsequent storms may trigger panic reactions. Immediate reactions to hurricanes can include emotional and physical exhaustion. In some instances children may experience survivor guilt (e.g., that they were not harmed, while others were killed or injured). Research indicates that greater symptomatology in children is associated with more frightening experiences during the storm and with greater levels of damage to their homes.

Earthquakes. Aftershocks differentiate earthquakes from other natural disasters. Since there is no clearly defined endpoint, the disruptions caused by continued tremors may increase psychological distress. Unlike other natural disasters (e.g., hurricanes and certain types of floods), earthquakes occur with virtually no warning. This fact limits the ability of disaster victims to make the psychological adjustments that can facilitate coping. This relative lack of predictability also significantly lessens feelings of controllability. While one can climb to higher ground during a flood, or install storm shutters before a hurricane, there is usually no advance warning or immediate preparation with earthquakes. Survivors may have to cope with reminders of the destruction (e.g., sounds of explosions, and the rumbling of aftershocks; smells of toxic fumes and smoke; and tastes of soot, rubber, and smoke).

Tornadoes. Like earthquakes, tornadoes can bring mass destruction in a matter of minutes, and individuals typically have little time to prepare. Confusion and frustration often follow. Similar to a hurricane, people experience sensations during tornadoes that may generate coping challenges. It can be difficult to cope with the sights and smells of destruction. Given the capricious nature of tornadoes, survivor guilt has been observed to be an especially common coping challenge. For instance, some children may express guilt that they still have a house to live in while their friend next door does not. In addition, a study following a tornado that caused considerable damage and loss of life revealed significant associations between children's disturbances and having been in the impact zone, been injured, and having experienced the death of relatives.

Floods. These events are one of the most common natural disasters. Flash floods are the most dangerous as they occur without warning; move at intense speeds; and can tear out trees, destroy roads and bridges, and wreck buildings. In cases of dam failure the water can be especially destructive. Research has reported that many children who survive a destructive flood experience psychological distress. The two most significant predictors of impairment are the degree of disaster exposure and perceptions of family reactions. Sensations that may generate coping challenges include desolation of the landscape, the smell of sludge and sodden property, coldness and wetness, and vast amounts of mud. Most floods do not recede overnight, and many residents have to wait days or weeks before they can begin the cleanup.

Most children will be able to cope over time with the help of parents and other caring adults.

Wildfires. Unlike other natural disasters such as earthquakes, there is often some warning of an advancing wildfire. However, depending upon the wind and terrain the direction and spread of a wildfire can change abruptly. The amount of warning can vary from one neighborhood to the next. While some people may have hours (or even days) to evacuate, others will have only a few minutes to gather their belongings and leave their homes. Even if evacuation is not ultimately necessary, preparing for the possibility can be frightening for children, particularly if they are seeing images of homes burning nearby on television.

Reactions immediately following a wildfire may include emotional and physical exhaustion. In some instances children may experience survivor guilt (e.g., that their home was left unharmed, while others were completely destroyed). In general it might be expected that greater symptomatology in children will be associated with more frightening experiences during the wildfire and with greater levels of damage to their community and homes. The sights, sounds, and smells of a wildfire often generate fear and anxiety. Consequently, similar sensations (e.g., the smell of smoke) may generate distress among children in the months that follow. Given the scale of most wildfires, individuals living outside the ravages of the fires may still feel exposed to the danger from drifting clouds of smoke, flames on the horizon, and television reports. Some children may also react to follow-up news coverage, and even weather reports that talk about dry fire conditions after the fact.

It is important to acknowledge that although a given natural disaster may last for only a short period, survivors can be involved with the disaster aftermath for months or even years. In attempts to reconstruct their lives following such a natural disaster, families are often required to deal with multiple people and agencies (e.g., insurance adjustors, contractors, electricians, roofers, the Red Cross, the Federal Emergency Management Agency, and the Salvation Army).

A minority of children may be at risk of post-traumatic stress disorder (PTSD).

Possible Reactions of Children and Youth

Most children will be able to cope over time with the help of parents and other caring adults. However, some children may be at risk of more extreme reactions. The severity of children's reactions will depend on their specific risk factors. These include exposure to the actual event, personal injury or loss of a loved one, dislocation from their home or community, level of parental support, the level of physical destruction, and pre-existing risks, such as a previous traumatic experience or mental illness. Symptoms may differ depending on age but can include:

- *Preschoolers*—thumb sucking, bedwetting, clinging to parents, sleep disturbances, loss of appetite, fear of the dark, regression in behavior, and withdrawal from friends and routines.

- *Elementary school children*—irritability, aggressiveness, clinginess, nightmares, school avoidance, poor concentration, and withdrawal from activities and friends.

- *Adolescents*—sleeping and eating disturbances, agitation, increase in conflicts, physical complaints, delinquent behavior, and poor concentration.

A minority of children may be at risk of post-traumatic stress disorder (PTSD). Symptoms can include those listed above, exhibited over an extended period of time. Other symptoms may include re-experiencing the disaster during play and/or dreams; anticipating or feeling that the disaster is happening again; avoiding reminders of the disaster; general numbness to emotional topics; and increased arousal symptoms such as inability to concentrate and startle reactions. Although rare, some adolescents may also be at increased risk of suicide if they suffer from serious mental health problems like PTSD or depression. Students who exhibit these symptoms should be referred for appropriate mental health evaluation and intervention.

Information for School Crisis Teams

Identify children and youth who are high risk and plan interventions. Risk factors are outlined in the above section on children's reactions. Interventions may include individual counseling, small group counseling, or family therapy. From group crisis interventions, and by maintaining close contact with teachers and parents, the school crisis response team can determine which students need supportive crisis intervention and counseling services. A mechanism also needs to be in place for self-referral and parental-referral of students.

Support teachers and other school staff. Provide staff members with information on the symptoms of children's stress reactions and guidance on how to handle class discussions and answer children's questions. As indicated, offer to help conduct a group discussion. Reinforce that teachers should pay attention to their own needs and not feel compelled to do anything they are not comfortable doing. Suggest that administrators provide time for staff to share their feelings and reactions on a voluntary basis as well as help staff develop support groups. In addition, teachers who had property damage or personal injury to themselves or family members will need leave time to attend to their needs.

The frequent need for disaster survivors to relocate creates unique crisis problems.

Engage in post-disaster activities that facilitate healing. [Pediatric psychologist A.M.] La Greca and colleagues have developed a manual for professionals working with elementary school children following a natural disaster. Activities in this manual emphasize three key components supported by the empirical literature: (a) exposure to discussion of disaster-related events, (b) promotion of positive coping and problem-solving skills, and (c) strengthening of children's friendship and peer support. Specifically:

- *Encourage children to talk about disaster-related events.* Children need an opportunity to discuss their experiences in a safe, accepting environment. Provide activities that enable children to discuss their experiences. These may include a range of methods (both verbal and nonverbal) and incorporate varying projects (e.g., drawing, stories, audio and video recording). Again provide teachers specific suggestions or offer to help with an activity.

- *Promote positive coping and problem-solving skills.* Activities should teach children how to apply problem-solving skills to disaster-related stressors. Children should be encouraged to develop realistic and positive methods of coping that increase their ability to manage their anxiety and to identify which strategies fit with each situation.

- *Strengthen children's friendship and peer support.* Children with strong emotional support from others are better able to cope with adversity. Children's relationships with peers can provide suggestions for how to cope with difficulties and can help decrease isolation. In many disaster situations, friendships may be disrupted because of family relocations. In some cases parents may be less available to provide support to their children because of their own distress and their feelings of being overwhelmed. It is important for children to develop supportive relationships with their teachers and classmates. Activities may include asking children to work cooperatively in small groups in order to enhance peer support.

Emphasize children's resiliency. Focus on their competencies in terms of their daily life and in other difficult times. Help children identify what they have done in the past that helped them cope when they were frightened or upset. Tell students about other communities that have experienced natural disasters and recovered (e.g., Miami, FL [Hurricane Andrew in 1992] and Charleston, SC [Hurricane Hugo in 1989]).

Parents' adjustment is an important factor in children's adjustment, and the adjustment of the child in turn contributes to the overall adjustment of the family.

Support all members of the crisis response team. All crisis response team members need an opportunity to process the crisis response. Providing crisis intervention is emotionally draining. This is likely to include teachers and other school staff if they have been serving as crisis caregivers for students.

Secure additional mental health support. Although more than enough caregivers are often willing to provide support during the immediate aftermath of a natural disaster, long-term services may be lacking. School psychologists and other school mental health professionals can help provide and coordinate mental health services, but it is important to connect with community resources in order to provide such long-term assistance. Ideally these relationships would be established in advance.

Important Influences on Coping

Relocation. The frequent need for disaster survivors to relocate creates unique crisis problems. For example, it may contribute to the social, environmental, and psychological stress experienced by disaster survivors. Research suggests that relocation is associated with higher levels of ecological stress, crowding, isolation, and social disruption.

Parents' reactions and family support. Parents' adjustment is an important factor in children's adjustment, and the adjustment of the child in turn contributes to the overall adjustment of the family. Altered family functions, separation from parents after natural disaster, and ongoing maternal preoccupation with the trauma are more predictive of trauma symptomatology in children than is the level of exposure. Thus, parents' reactions and family support following a natural disaster are important considerations in helping children cope.

Emotional reactivity. Preliminary findings suggest that children who tend to be anxious are those most likely to develop post-trauma symptomatology following a natural disaster. Research suggests that children who had a preexisting anxiety disorder prior to a natural disaster are at greater risk of developing PTSD symptoms.

Coping style. It is important to examine children's coping following a natural disaster because coping responses appear to influence the process of adapting to traumatic events. Research suggests that the use of blame and anger as a way of coping may create more distress for children following disasters.

Long-Term Effects

Research suggests that long-term difficulties following a natural disaster (e.g., PTSD), are most likely to be seem among children who experienced any of the following:

- Had threats to their physical safety.

- Thought they might die during the disaster.

- Report that they were very upset during the disaster.

- Lost their belongings or house as a result of the disaster.

- Had to relocate in the aftermath.

- Attended schools following the disaster that had multiple schedule changes, double sessions or a lot of disruptions.

Consequently, crisis response team members need to identify students who experience these risk factors and closely monitor their status. These students may require long-term coping assistance.

Disaster Response Challenges Are Mounting as Global Warming Increases Climatic Disasters Around the World

Oxfam International

Oxfam International is an international relief and development organization that seeks lasting solutions to poverty, hunger, and injustice.

Climatic disasters are on the increase as the Earth warms up—in line with scientific observations and computer simulations that model future climate. 2007 [was] a year of climatic crises, especially floods, often of an unprecedented nature. They included Africa's worst floods in three decades, unprecedented flooding in Mexico, massive floods in South Asia and heat waves and forest fires in Europe, Australia, and California. By mid November the United Nations had launched 15 'flash appeals', the greatest ever number in one year. All but one were in response to climatic disasters.

The Vulnerability of the Poor

At the same time as climate hazards are growing in number, more people are being affected by them because of poverty, powerlessness, population growth, and the movement and displacement of people to marginal areas. The total number of natural disasters has quadrupled in the last two decades— most of them floods, cyclones, and storms. Over the same period the number of people affected by disasters has increased from around 174 million to an average of over 250 million a year. Small- and medium-scale disasters are occurring more frequently than the kind of large-scale disasters that hit the headlines.

However, dramatic weather events do not in themselves necessarily constitute disasters; that depends on the level of human vulnerability—the capacity to resist impacts. Poor people and countries are far more vulnerable because of their poverty. Disasters, in turn, undermine development that can provide greater resilience.

Accelerating climate change is bringing more floods, droughts, extreme weather and unpredictable seasons.

One shock after another, even if each is fairly small, can push poor people and communities into a downward spiral of destitution and further vulnerability from which they struggle to recover. Such shocks can be weather-related, due to economic downturns, or occur because of conflict or the spread of diseases like HIV and AIDS. Such shocks hit women hardest; they are the main collectors of water and depend most directly on access to natural resources to feed their families; they have fewer assets than men to fall back on, and often less power to demand their rights to protection and assistance.

Now, accelerating climate change is bringing more floods, droughts, extreme weather and unpredictable seasons. Climate change has the potential to massively increase global poverty and inequality, punishing first, and most, the very people least responsible for greenhouse-gas emissions—and increasing their vulnerability to disaster.

Hope for the Future

There is hope. The global humanitarian system has been getting better at reducing death rates from public-health crises following on from major shocks like floods or droughts. But humanitarian response is still skewed . . . to high-profile disasters, and it will certainly be woefully inadequate as global temperatures continue to rise, unless action is taken quickly.

The world has an immediate responsibility to stem the increase in climate-related hazards.

New approaches to humanitarian action are needed as well as new money. Political efforts aimed at reducing poverty and inequality, which provide people with essential services like health and education, and offer social protection (a regular basic income, or insurance), constitute a firm foundation for effective disaster risk reduction (DRR), preparedness, and response. More work needs to be done to understand the linkages between development, DRR and climate change adaptation, and therefore to more accurately assess the financial costs climate change will impose.

Fundamentally, the world has an immediate responsibility to stem the increase in climate-related hazards. Above all, that means tackling climate change by drastically reducing greenhouse-gas emissions.

Oxfam's recommendations are:

Mitigate: Greenhouse-gas emissions must be reduced drastically to keep global average temperature rise as far below two degrees Celsius as possible. Rich countries must act first and fastest. . . .

Adapt: Oxfam has estimated that, in addition to funding for emergency response, developing countries will require at least US$50bn [billion] annually to adapt to unavoidable climate change. These funds should be provided by rich nations in line with their responsibility for causing climate change and their capacity to assist. Additional finance for adaptation is not aid, but a form of compensatory finance; it must not come out of long-standing donor commitments to provide 0.7 per cent of gross domestic product [a measure of a country's total economic output] as aid in order to eradicate poverty. At present, funding for adaptation is totally inadequate, and the [2007] UN ... must mandate the search for new funds. Innovative financing mechanisms need to be explored.

Improve the global humanitarian system:

- *Increase emergency aid*: Major donor governments must keep their promises to increase Overseas Development Aid by an additional US$50bn a year by 2010. If they do, then humanitarian aid is likely to increase in proportion from over US$8bn to over US$11bn. But aid is going in the wrong direction, and anyway this is unlikely to be enough; increased warming and climate change pose the very real danger that humanitarian response will be overwhelmed in the coming decades. More money and better responses are both needed.

- *Ensure fast, fair, flexible, appropriate aid*: This should include moving away from over-reliance on in-kind food aid, towards more flexible solutions such as cash transfers.

Reduce vulnerability and the risk of disaster:

- *Build long-term social protection*: Climate change is accentuating the fact that for many poor people, shocks are the norm. Governments must put poor people first. Aid should be used to build and protect the livelihoods and assets of poor people. Providing essential services like water, sanitation, health and education, and long-term social protection systems form the foundations for timely emergency scale-up when required.

- *Invest in disaster risk reduction (DRR)*: Governments have made commitments to make the world safer from natural hazards through investing in DRR approaches. They need to put their promises into action and link DRR to both climate-change adaptation measures and to national poverty reduction strategies.

- *Build local capacity*: Build the capacity of local actors, particularly government at all levels, and empower affected populations so they have a strong role and voice in preparedness, response and subsequent recovery and rehabilitation.

- *Do development right*: Development aid should integrate analyses of disaster risk and climate trends. Inappropriate development strategies not only waste scant resources, they also end up putting more people at risk, for instance by the current reckless rush to produce biofuels without adequate safeguards for poor people and important environments.

Building a Disaster-Resilient Nation and World Is the Grand Challenge

David Applegate

David Applegate is chair of the Subcommittee on Disaster Reduction, part of the National Science and Technology Council, which advises the president on science and technology issues. He is also the senior science adviser for earthquake and geologic hazards at the U.S. Geological Survey, part of the U.S. Department of the Interior.

The deadly typhoon that struck Burma [also called Myanmar] in early May [2008] and the devastating earthquake that struck China a week later carried with them echoes of the devastation wrought by the Sumatra earthquake and tsunami on the Indian Ocean region in 2004. In the United States, the vulnerability of Burma's coastal populations to severe winds and storm-surge inundation also served to remind us of Hurricane Katrina and the ongoing recovery in New Orleans and other Gulf Coast communities. Although U.S. cities have not experienced a catastrophic earthquake since the one in Anchorage, Alaska, in 1964, we know that events as large as the Chinese earthquake will strike in the future.

Becoming Resilient

Extreme events are facts of life on an active planet like ours. How those events affect us reflects not only the power of nature but the decisions we make in how we build our societies. Achieving security at home and abroad must reflect an overall resilience to all hazards that confront our communities.

Achieving that resilience is a grand challenge, and it will take the collective action of government at all levels, nonprofit organizations, the private sector and above all individuals trying to do what is best for themselves, their families and their communities.

Science and technology can play a critical role in the quest for disaster resilience. To better define this role, the National Science and Technology Council's Subcommittee on Disaster Reduction, representing 22 federal departments and agencies, identified six grand challenges for disaster reduction.

The Grand Challenges

The first of these challenges is to provide hazard and disaster information where and when it is needed. Meeting this challenge requires robust monitoring systems with the capability to reach those in harm's way and provide emergency responders with the information they need. Such systems are only as good as their weakest link, which in many cases is the link to the people at risk. Improving communications to the most vulnerable populations, so that they can protect themselves, requires education.

> *Achieving security at home and abroad must reflect an overall resilience to all hazards that confront our communities.*

The second challenge is to understand the natural processes that produce hazards. Targeted research can harness advances in computing power and draw upon data generated by global observational systems to improve predictive modeling. For coastal hazards, this understanding must include assessment of the impacts of climate change on coastal inundation.

The third challenge is to develop strategies and technologies to reduce the impact of extreme events on the built environment and vulnerable ecosystems. Meeting this challenge will require understanding social, cultural and economic factors that promote or inhibit promising mitigation technologies.

The fourth grand challenge is to reduce the vulnerability of infrastructure. One major obstacle to recovery in any disaster is the delayed restoration of critical infrastructure such as drinking water, electricity and gas distribution systems. A key step is establishing the technical basis for revised codes and standards for critical infrastructure. Paradoxically, advances in technology can increase society's vulnerability because of reliance on distant resources and just-in-time inventory delivery, with the result that the economic impact of a natural hazard event can be broader than its storm track or rupture zone.

The fifth challenge is to develop standardized methods for communities to measure and assess disaster resilience across multiple hazards. A key step is developing and distributing assessment tools that can be used to set priorities.

The final challenge is to promote risk-wise behavior. The costs of natural disasters are rising as people increasingly move into harm's way in low-lying coastal areas, the wildland-urban interface and geologically active regions. In order to achieve "hazards literacy" and sustained risk reduction, hazards must be real to people. Scenarios are a tool that can spell out the impacts of likely events on high-risk areas, combining scientific and engineering knowledge with local planning and emergency management expertise to deliver a comprehensive picture of potential losses.

The subcommittee has released plans identifying the priority science and technology actions needed to meet these challenges for all major hazards. These plans will help to shape sustained federal science and technology investments in disaster reduction and can also serve as a blueprint for international cooperation.

All the plans identify the same desired outcomes: A nation where relevant hazards are recognized and understood, communities at risk know when a hazard event is imminent, individuals can live safely in the context of our planet's extreme events and communities experience minimum disruption to life and economy after a hazard event.

Current
CONTROVERSIES

CHAPTER 2

Has the U.S. Disaster Response Improved Since Hurricane Katrina?

Chapter Preface

The widespread criticism of the federal, state, and local response to Hurricane Katrina in 2005 quickly led to calls for an official investigation to determine what went wrong and how it could be fixed. Shortly after the disaster, on September 7, 2005, Republicans—who controlled both houses of Congress—announced a joint House of Representatives/Senate investigation, but this proposal was dropped soon thereafter. Instead, the House of Representatives on September 15, 2005, formed what was called the Select Bipartisan Committee to Investigate the Preparation for the Response to Hurricane Katrina.

Congressional Democrats, however, refused to participate in the House investigation because they believed that it was not truly bipartisan and would be used by Republicans to limit political damage to their party from the disaster. Democrats, led by Senator Hillary Clinton, called instead for an independent commission modeled on the 9-11 Commission, which was formed to investigate the September 11, 2001, terrorist attacks. But on September 14, 2005 Senate Republicans voted down an amendment offered by Senator Clinton to a spending bill that would have created the independent commission. A few months later, on February 2, 2006, Senate Republicans again voted to kill a second effort by Senator Clinton to create such a commission. In lieu of the independent commission, Senate Republicans chose to support an investigation by the Senate Homeland Security and Governmental Affairs Committee, chaired by Democratic senator Joe Lieberman. The Senate's investigation was launched by this committee on September 15, 2005.

President George W. Bush promised that the White House would support congressional inquiries and also conduct its own investigation of the federal, state, and local response to

Katrina. By most accounts, however, the Bush administration hindered the investigation process and sought to downplay its responsibility for the poor disaster response to Katrina. Although some administration officials were permitted to testify and some documents were provided to congressional committees conducting investigations into the Katrina debacle, the Bush administration refused to turn over some important documents about Hurricane Katrina and declined to make senior White House officials available for sworn testimony before Congress.

The White House cited the confidentiality of executive branch communications as the reason for its actions, but members of Congress, both Democrats and Republicans, claimed President Bush was stonewalling the committees and making it impossible for them to conduct thorough investigations. In fact, both committees ultimately failed to acquire documents and testimony that many commentators believed were essential to investigating the White House role in the disaster. Examples of missing materials included a videoconference in the White House in which former Federal Emergency Management Agency (FEMA) director Michael Brown said he warned senior officials about the dire situation in New Orleans, but was greeted with "deafening silence," as well as messages and conversations involving the president, vice president, and their top aides as the Katrina disaster was unfolding.

In addition, the administration's own investigation produced a report in February 2006 that was widely viewed as a cover-up of the federal-level failures. The report, entitled "The Federal Response to Hurricane Katrina: Lessons Learned," stated that it sought to avoid blame, praised the administration's efforts to protect people and property, and failed to hold FEMA, the federal agency that received the most criticism in the aftermath of Katrina, accountable for its actions.

Despite the lack of White House cooperation, both the Senate and House issued reports on the Katrina disaster. The House of Representatives' report, called "A Failure of Initiative," identified failures at all levels of government, said FEMA and the White House were overwhelmed, and found that the nation was not fully prepared to respond to either natural or human-caused disasters. The Senate likewise found that FEMA was unprepared and found leadership failures at all levels of government, including the White House. The Senate recommendations included, among others, abolishing FEMA and replacing it with a new agency; improving coordination between federal, state, and local governments; and increasing resources for disaster preparation and response.

In the fall of 2006, Congress enacted the Post-Katrina Emergency Reform Act (PKEMRA) to reorganize FEMA and improve its effectiveness in responding to disasters. In April 2007, the government created a new FEMA, which is supposed to be better able to coordinate federal functions, work with state and local entities, and respond quickly in the event of a major disaster. Some commentators have praised the new agency and reported favorably on its response to several post-Katrina disasters, but others maintain that the United States disaster response system is still inadequate. The viewpoints included in this chapter address this critical question of whether the reforms instituted since Hurricane Katrina have improved the country's disaster response system.

A New and Enhanced FEMA Was Created in 2007

R. David Paulison

R. David Paulison was the director of the Federal Emergency Management Agency (FEMA) from 2005 to 2008. FEMA is the main federal agency in the United States responsible for disaster preparedness and response.

The federal response to the 2005 hurricanes was a clarion call for change in disaster response and recovery for the country and all of those involved in emergency management. Based on the many lessons learned, FEMA [Federal Emergency Management Agency] instituted numerous reforms to improve its ability to respond to and recover from disasters. In addition to FEMA's internal transformation that we embraced to improve the agency, the Department of Homeland Security (DHS) and FEMA have been working closely with other components within DHS to implement the adjustments included in the Post-Katrina Emergency Management Reform Act (PKEMRA or "the Act"). The combination of FEMA's transformation and changes made by PKEMRA are resulting in a new FEMA that is stronger, more nimble and more robust than we were just a year ago [in mid-2006]. . . .

The New FEMA—Organization

Last fall [2006], Congress passed and the President signed into law the FY [fiscal year] 2007 Homeland Security Appropriations Act, which included PKEMRA. The legislation reorganizes DHS and reconfigures FEMA to include consolidated emergency management functions, including national preparedness functions.

R. David Paulison, testimony before the U.S. House of Representatives Committee on Oversight and Government Reform, Washington, D.C., July 31, 2007. Reproduced by permission. http://oversight.house.gov/documents/20070731105123.pdf.

Significantly, and consistent with the lessons learned, the new FEMA has not simply tacked on new programs and responsibilities to an existing structure. Rather, we conducted a thorough assessment of the internal FEMA structure, including new and existing competencies and responsibilities within FEMA. On April 1 [2007], this new and expanded FEMA was formally established. This new organization reflects the expanded scope of FEMA's responsibilities—and the core competencies that we are seeking to establish and enhance. It supports a more nimble, flexible use of resources. It strengthens coordination among FEMA elements and with other DHS components. It enables FEMA to better coordinate with agencies and departments outside of DHS. It also delivers enhanced capabilities to our partners at the state, local and tribal governments and emergency management and preparedness organizations at all levels, and engages the capabilities and strengths that reside in the private sector. . . .

Of particular note in the reorganization, the new FEMA includes a new National Preparedness Directorate, which incorporates functions related to preparedness doctrine, policy and contingency planning. It also contains the exercise coordination and evaluation program, emergency management training, the Chemical Stockpile Emergency Preparedness Program and the Radiological Emergency Preparedness program.

The new FEMA has not simply tacked on new programs and responsibilities to an existing structure.

In addition to preparedness, the new FEMA is sharpening the agency's focus on building core competencies in logistics, operational planning, incident management and the delivery of disaster assistance. . . .

Where the "rubber really meets the road" is in FEMA's regional offices. Having key leaders with the necessary experience and adequate resources to support their missions across the country is an important element of the agency's reorganization. The ten Regional Administrators report directly to me, the Administrator [as this agency calls its director], and are supported by Regional Advisory Councils. The Regional Advisory Councils provide advice and recommendations to the Regional Administrators on regional emergency management issues and identify weaknesses or deficiencies in preparedness, protection, response, recovery and mitigation for state, local and tribal governments based on their specialized knowledge of the region. We have filled all 10 Regional Administrator posts with men and women with 20 to 30 years of emergency management experience. We also are working to improve operational capabilities in the regions and will establish Incident Management Assist Teams (IMATs) in them. The IMATs will support the enhanced regions by providing a dedicated 24/7 organic response capability. When not deployed, IMATs will train with and enhance the emergency management capabilities of our federal, state, local and tribal partners. . . .

The result of these changes is a new FEMA that is better prepared for the future than the organization was in the past.

The New FEMA—Planning and Operations

The result of these changes is a new FEMA that is better prepared for the future than the organization was in the past. We have a stronger organization with stronger leaders and dedicated men and women striving to serve those most in need. But these structural changes will not be meaningful unless matched with a similar change in FEMA's vision and goals. We are working diligently to reestablish America's trust and confidence in FEMA, and are focused on our vision to become the nation's preeminent emergency management and preparedness agency.

The guiding principle of this new FEMA is to lean further forward to deliver more effective disaster assistance to individuals and communities impacted by a disaster. We call it "engaged partnership." This partnership was evident in the Florida, Georgia, and Alabama tornadoes, the Nor'easter that affected the New England states, and in Kansas, where the community of Greensburg was devastated by a tornado.

In these disasters, FEMA was engaged with the state within minutes of the disaster, immediately deployed operational and technical experts to the disaster site, started moving logistics and communications capabilities even before a disaster declaration and coordinated with the Governor to facilitate a Presidential disaster declaration. Also, FEMA has supported and helped to facilitate an effective Unified Command with other federal agencies, and state and local officials.

FEMA's support of the response operations for states impacted by large, uncontrolled wildfires is a prime example of our ability to lean forward. FEMA provides Fire Management Assistance Grants (FMAGs) to states when a fire threatens such destruction as would constitute a major disaster. FMAG declarations operate on a 24-hour, real time basis to provide assistance through emergency protective measures, which may include grants, equipment, supplies, and personnel for the mitigation and management of a fire threatening a major disaster. This year [2007], FEMA has provided assistance in the form of FMAG declarations for over 30 fires across 13 states. This assistance includes support of the states of California, Florida, Georgia, and Utah, all of which have experienced extreme fire activity this year. California received seven FMAGs, including one in support of the Angora Fire which significantly threatened the communities around Lake Tahoe. The states of Florida and Georgia received a combined total of 11 FMAGs to assist with their unprecedented early fire season. Utah received assistance for two of the largest fires in the state's history. These grants were declared within hours of being requested, often in the middle of the night, to make available federal assistance to protect citizens and critical facilities. These efforts demonstrate FEMA's ability to support and help facilitate an effective unified response with other federal agencies, and state and local officials.

All of these actions were taken by a well-led, motivated, and professional FEMA workforce that has embraced and enhanced the vision and reality of a new FEMA. . . .

Heightened Hurricane Preparedness

FEMA is placing its primary emphasis on strengthening the federal-state partnership to better ensure that we are able to achieve shared objectives for a safe, coordinated and effective response and recovery effort [for future hurricanes].

The guiding principle of this new FEMA is to lean further forward to deliver more effective disaster assistance to individuals and communities impacted by a disaster.

First, we are emphasizing the states' primary responsibility to provide for the safety and security of their citizens. The states must take the lead to ensure they and their local jurisdictions are prepared for the hurricane season.

The various state emergency management agencies coordinate the overall management of an emergency to include requests for support and resources from other state agencies, from other states under the Emergency Management Assistance Compact (EMAC), and for supplemental assistance from the federal government. The EMAC process offers state-to-state assistance during Governor-declared states of emergency. EMAC offers a responsive and straightforward system under which states can send personnel and equipment to help disaster relief efforts in other states. When one state's resources are overwhelmed, other states can help to fill the shortfalls through EMAC.

The strength of EMAC and the quality that distinguishes it from other plans and compacts lies in its governance structure, its relationship with federal organizations, states, counties, territories, and regions, and the ability to move just about any resource from one state to another.

Secondly, by advancing the concept of engaged partnership, FEMA will stand shoulder-to-shoulder with the state—we are there to support, fill gaps, and help to achieve a successful response and recovery. In the past, our system was cued to sequential failure: where the state held back until the local jurisdiction was overwhelmed, and the federal system held back until the state was overwhelmed. This approach, evident in the response to Katrina, caused delays in delivering support. Under "engaged partnership," FEMA has strengthened the relationship between FEMA Regional Administrators and State Emergency Managers to focus on more deliberate disaster planning. . . .

Third, FEMA has extended its reach across the span of federal agencies to ensure the smooth and responsive coordination of federal support when it is needed. The most visible demonstration of that coordination is the array of federal capabilities contained in our "playbook" of pre-scripted mission assignments. This playbook represents an examination of the range of federal support that may be requested in response to a disaster. It also includes advance interagency coordination to ensure delivery of that capability when called upon in time of need. At present, we have developed and coordinated 187 pre-scripted mission assignments with as many as 21 federal agencies. Up to an additional 40 are still under review. This support ranges from heavy-lift helicopters from the Department of Defense (DOD), to generators from the U.S. Army Corps of Engineers, to Disaster Medical Assistance Teams from Health and Human Services (HHS) and Emergency Road Clearing Teams from the U.S. Forest Service. These pre-scripted mission assignments will result in more rapid and responsive delivery of federal support.

I believe we have made real progress at FEMA and are prepared for [future disasters]. . . . Our reorganization efforts—based on our internal transformation and the implementation of PKEMRA—will bear fruit across our disaster operations and assistance programs. Today, I have focused on how FEMA's reorganization has contributed to:

1. Establishing a heightened posture of hurricane preparedness;

2. Engaging our state and federal partners in more thorough and informed hurricane planning; and,

3. Building FEMA's operational capabilities to provide effective response and recovery.

There is a lot more going on inside FEMA than the things mentioned that will also contribute to enhanced performance and organizational success.

Although all disasters are local, FEMA must play a more proactive role in understanding vulnerabilities so we can assist the localities in being better prepared to respond. As I hope you will see by today's testimony—we are being more proactive. By leaning further forward to coordinate the federal response, we can better serve all Americans.

The Federal Government Has Improved Disaster Preparedness and Response

Gregg Carlstrom

Gregg Carlstrom is a reporter for Federal Times, *a trade magazine for federal government employees.*

It hasn't been business as usual for federal agencies responding to this summer's [2008] wave of natural disasters.

After the floods in the Midwest, Health and Human Services Department [HHS] officials quickly authorized new daycare centers for more than 2,500 displaced children. The Agriculture Department's Food and Nutrition Service sped up food assistance to more than 12,000 displaced Iowans.

And during the California wildfires, the Federal Emergency Management Agency [FEMA] trained 2,000 California National Guard members as firefighters, helping to relieve the state's own beleaguered fire departments. Since Hurricane Katrina, agencies like HHS and FEMA say a stronger focus on risk analysis and a better partnership with states and municipalities have improved disaster preparedness and response.

"We're getting more analytical based on what we've learned from previous events," said Dr. Kevin Yeskey, HHS' deputy assistant secretary for preparedness and response. "We have a [team] analyzing our performance, and we're doing after-action reports . . . to look at where things didn't happen" and how to improve.

Coordinating with States

Much of HHS's disaster recovery role is focused on coordinating with states and municipalities. The department has emergency coordinators in each of the country's 10 regions. They develop relationships with local officials and take the lead on analyzing the capabilities of local governments.

The federal government often doesn't have the resources to directly help in a disaster area, so it facilitates help from parties that do, Yeskey said.

Since Hurricane Katrina, . . . a stronger focus on risk analysis and a better partnership with states and municipalities have improved disaster preparedness and response.

"After the Midwest flooding, Iowa needed a mobile hospital unit," he said, offering one example. "We don't have those. Municipalities have those. And [Iowa] was able to get one from North Carolina . . . without federal interaction, through a partnership we helped facilitate."

FEMA is similarly focused on coordinating with states, according to Glenn Cannon, assistant administrator for disaster operations. Just last week [in August 2008], Cannon said, FEMA officials were monitoring two Atlantic storm systems and holding conference calls with officials in coastal states and offshore territories. "We often don't have a lot of resources," he said. "But we coordinate the response."

That coordination requires lots of employees in the field. FEMA has gone on a hiring binge since Katrina—by next year, Cannon estimates, staffing could be double the pre-Katrina level—and most of those employees are based in field offices.

The same is true at HHS, where regional employees take the lead on developing relationships with local governments.

"We spend most of our time building relationships, so people are familiar with us, and so we know what they're capable of," said Fred Shuster, HHS regional director in Kansas City.

Dana Hall, a regional emergency coordinator for HHS in Kansas City, cited one example of this: She recently traveled to North Dakota to strike an agreement among nine Midwestern states to share their laboratory resources in case of a pandemic or a biological attack.

Improving Disaster Response

There's also been an effort to improve the response from state and local governments, and the private sector. HHS advised hospitals in several major cities to take small steps—like moving generators out of the basement, where they could be flooded, and onto higher floors—that could save money after a disaster.

And FEMA has conducted gap analyses of the 18 coastal states that are prone to hurricanes. These studies look for potential shortfalls—not having enough hospital capacity, for example, or a shortage of buses for evacuation—and then propose remedies. To respond to those studies, FEMA has pre-staged assets from Maine to Florida, and along the Gulf Coast, to facilitate a quicker response.

Emergency management officials in Iowa and Illinois—states hit hard [in 2008] by flooding—had praise for the government's quick response.

A key element of those analyses is what FEMA and HHS call "prescripted mission assignments." These are essentially checklists for federal, state and local partners: lists of required actions, necessary supplies, and instructions for getting supplies to the disaster site. FEMA had 17 of those checklists before Katrina; today, the agency has more than 200, many of them for the Centers for Disease Control and Prevention [CDC], the Army Corps of Engineers and the Pentagon.

"We want to know what their vulnerabilities are, and what we need to have in place," Cannon said. "When something bad is happening, that's not the time to figure out what you need."

Improving Communications

HHS has also taken a different approach to its information flow. The department set up a 24-hour command center at its Southwest Washington headquarters. On a Friday afternoon visit, the center was quiet—staffed by about 10 employees—but Yeskey said that number swells to several dozen during an incident.

The command center acts as a conduit, managing the flow of information between headquarters and field offices. The department also brings in liaisons from other agencies with disaster responsibilities, like FEMA, the Homeland Security Department and CDC.

FEMA has gone a step further, opening command centers in each of its 10 regions. Half of them are operating continuously, and the agency expects every center to be 24/7 by the end of 2009. At the state level, emergency officials tend to agree that agencies like FEMA and HHS are doing a better job of responding to disasters.

The pace of change since Katrina has been remarkable . . . and a break from the less-responsive, reactive agencies of the past.

Emergency management officials in Iowa and Illinois—states hit hard by flooding—had praise for the government's quick response, particularly FEMA's efforts to shelter displaced residents.

Increased Staffing

All of these changes have meant big changes to budgets and staffing levels at the most active disaster-response agencies.

At FEMA, staffing has increased by more than 400 employees [in 2008]; FEMA's Cannon estimates that employment could be 3,400 by next year, double the pre-Katrina levels.

At HHS, staffing levels have also increased: from 160 employees working on disaster operations to more than 240. But the department's disaster budget has decreased, from $554 million to $486 million, more than 12 percent. Most of the decrease is in the Hospital Preparedness Program; HHS officials said their gap analyses allow them to better target their funds, so money is spent only where it's needed. The increases for staffing and new technology more than offset the decrease in the hospital fund.

Officials at both agencies say the pace of change since Katrina has been remarkable. But they also describe the changes as long overdue—and a break from the less-responsive, reactive agencies of the past.

"Before Katrina, we were essentially a coordinating, check-writing agency, but the public and Congress had an expectation beyond that," Cannon said of FEMA.

Coordination Between the Federal Government and States Was Better for Hurricane Gustav than for Hurricane Katrina

Associated Press

The Associated Press is an American news agency and wire service cooperatively owned by more than seventeen hundred newspapers and five thousand television and radio broadcasters.

President [George W.] Bush said Monday [September 1, 2008,] that coordination among states and the federal government in response to Hurricane Gustav has been better than during Katrina, which devastated the Gulf Coast in 2005 and tattered his administration's reputation for handling crises.

Bush visited an emergency operations center in Austin, [Texas,] about 400 miles west of Cocodrie, La., near where Gustav struck land. Gustav packed more than 100 mph winds, but delivered only a glancing blow to New Orleans, raising hopes that the city would escape the kind of catastrophic flooding wrought by Katrina, which killed nearly 1,600, obliterated 90,000 square miles of property and cost billions of dollars in response and repairs.

At an emergency operations center in Austin, Texas, Bush said the federal government's job was to assist states affected by the storm. He said he wanted to ensure that assets were in place to handle the storm, and that preparations are being made to help the Gulf Coast recover.

"The coordination on this storm is a lot better than . . . during Katrina," Bush said noting how the governors of Alabama, Louisiana, Mississippi and Texas had been working in concert. "It was clearly a spirit of sharing assets, of listening to somebody's problems and saying, 'How can we best address them?'"

Coordination among states and the federal government in response to Hurricane Gustav has been better than during Katrina.

Thanks to the Evacuees

He lauded Gulf Coast residents who heeded warnings to evacuate.

"It's hard for a citizen to pull up stakes, and move out of their home, and face the uncertainty that comes when you're not at home, and I want to thank those citizens who listened carefully to their local authorities and evacuated," Bush said.

Later, at another command center in San Antonio, [Texas,] Bush made a plea for Americans to help support recovery efforts by donating to relief agencies.

"Nobody's happy about these storms," he said. "Everybody's praying for everybody's safety, but I'm confident that after the storm passes, if there's a human need, it will be met because of the generosity of the American people."

David Paulison, director of the Federal Emergency Management Agency, told reporters on Bush's plane en route that there has been "unprecedented cooperation" among federal agencies and the private sector. "What it allows us to do is share information of what's going on so we don't end up with what happened in Katrina, with different agencies doing things and others not knowing what's happening," he said.

Paulison said the help came ahead of the storm time, significantly easing evacuations. Everyone in New Orleans who wanted to evacuate could have, Paulison said. "There should not be any excuses," he said. "If people stayed in New Orleans, it was their choice."

The enduring memory of Katrina is not the ferocity of the storm, but the bungled reaction that led to preventable deaths and chaos. Disaster response has undoubtedly improved since then. But Katrina was a low chapter in American history, and it deeply eroded credibility in Bush's administration.

Lessons Learned

Texas Gov. Rick Perry and Sen. Kay Bailey Hutchison, R-Texas, greeted Bush as he got off the plane in shirt sleeves on a hot, sunny day in Texas.

Perry said state authorities had evacuated 10,000 special needs citizens from the Texas coast and about 280,000 other Texans have been evacuated from Orange, Jefferson and Chambers counties. In all, Texas anticipates 45,000 to 50,000 evacuees from Louisiana, Perry said. During the briefing, Perry turned to Bush and said, "Your home state did good."

The lessons that were learned from Katrina can serve the United States very well in any kind of disaster.

By flying to Texas, Bush clearly wanted to show the nation, and particularly people of the Gulf Coast, that he is committed to answering their needs. He said he hopes to get to Louisiana, too, but will choose a time that does not interfere with emergency response efforts.

First lady Laura Bush also was involved in the administration's effort to stress that things would be different this time. "Mistakes were made by everyone" at all levels of government in the handling of Katrina, Mrs. Bush said Monday on CNN.

"Part of it was not being able to have the good communication that you would need between the three governments," said Mrs. Bush, who also was to speak Monday at the GOP convention. "And we have taken care of that, we know that's a lot better. And the lessons that were learned from Katrina can serve the United States very well in any kind of disaster."

Hurricanes Gustav and Ike Showed That FEMA Reforms Have Worked

Jeff Greene

Jeff Greene is a deputy director of the Project on National Security Reform. He was formerly counsel to the Senate's special investigation into Hurricane Katrina.

While the final cost of Hurricanes Gustav and Ike may not be known for weeks or even months, one thing is clear: The federal government has learned some of the harsh lessons taught by Hurricane Katrina in 2005.

In contrast to their poor performance three years ago [2005], the Department of Homeland Security [DHS] and the Federal Emergency Management Agency [FEMA] actively led the federal government's pre-storm preparations for Gustav and Ike [in 2008]. Agencies that usually fight over turf coordinated their activities, and federal officials relied on local expertise and guidance.

The Failure of Katrina

In the three years since Hurricane Katrina, I have observed these improvements from several vantage points: first as counsel to the Senate's investigation into Katrina and later as a subcommittee staff director with the House Committee on Homeland Security.

Today I work for the Project on National Security Reform [PNSR], a non-profit and non-partisan group that is studying how the federal government can better prepare for and respond to 21st century national security threats.

PNSR's recently issued Preliminary Findings Report revealed that an underlying problem that led to the Katrina disaster—the federal government's failure to coordinate efforts across agency lines and the competition between agencies looking out for their own bureaucratic interests—is widespread and detrimental to our ability to deal with unpredictable national security challenges. This includes natural disasters.

> *The federal government's failure to coordinate efforts across agency lines and the competition between agencies . . . is widespread and detrimental to our ability to deal with unpredictable national security challenges.*

PNSR will recommend steps . . . to improve our Cold War–era national security system. As Congress and the next president consider these badly needed changes, the post-Katrina reform process merits study.

In the aftermath of Katrina, the public's ire was squarely focused on FEMA. A lesser discussed but perhaps more insidious problem was the failings that occurred above FEMA, at DHS and the other agencies that did little to prepare as the hurricane approached and failed to engage until days after the storm.

The Senate's bipartisan Katrina investigation uncovered a federal government that was strangely dormant as Katrina approached. Evacuation assistance was not contemplated, no new supplies were pre-positioned, and agencies failed to coordinate about what could and should be done after the storm passed.

The problems exposed by Katrina were extensively addressed through congressional and executive branch reform efforts.

Two major pieces of legislation—the Post Katrina Emergency Management Reform Act of 2006 and the Implementing the Sept. 11 Commission Recommendations Act of 2007—contained provisions designed to get the federal government to coordinate better both before and after a disaster. The executive branch conducted its own lessons-learned effort and significantly revamped its procedures as well.

The Response to Hurricanes Ike and Gustav

Hurricanes Ike and Gustav were major tests of these changes. As both storms approached, FEMA and DHS pre-positioned an enormous amount of supplies—hundreds of generators; truckloads of tarps, cots and blankets; and millions of meals and cases of water.

And while the relief efforts after Ike and Gustav were by no means perfect, the intense pre-storm effort prevented a repeat of the widespread misery seen after Katrina. This is in no small part because the thousands of federal, state and local responders on the front lines facing Ike and Gustav were provided with the leadership and support that was lacking in 2005.

Why did post-Katrina reform work when so many other efforts have not? And what can we learn from this experience as we strive to remake our national security system to face the threats of the 21st century?

First, the effort worked because Congress put policy before partisanship and recognized the deadly serious need to improve preparation and response for future disasters.

As [Hurricanes Gustav and Ike] approached, FEMA and DHS pre-positioned an enormous amount of supplies.

And while the leadership of the executive branch was reluctant to acknowledge error after Katrina and initially resistant to congressionally mandated changes, rank-and-file employees were not. They were out ahead of some of their leaders, figuring out ways to do better next time.

Second, before Katrina both our government and many of our citizens simply could not conceive of destruction of such magnitude from a natural disaster. The attacks of Sept. 11, 2001, had, quite naturally, led to a myopic view of the threats we face. A key part of the reform effort was changing the way we think about security.

Third, while the reflexive reaction to Katrina was to call for yet another massive structural reorganization, in the end the post-Katrina reformers took a more measured approach.

Significant changes were made where needed—for instance, folding the DHS Preparedness Directorate back into FEMA and strengthening FEMA's regional structures. Elsewhere, though, incremental changes resulted in operational improvements, such as creating processes to facilitate interagency coordination.

The successful post-Katrina reforms—in particular, the process that led to them—are an example that can and should be followed as the United States considers the future of our national security system.

FEMA Reforms Did Not Address the Root Causes of the Katrina Response Disaster

Peter Grier

Peter Grier is a staff reporter for The Christian Science Monitor, *an international newspaper.*

A new response force of 1,500 full-time employees. Satellite tracking of trucks carrying food, bedding, and other relief supplies. Reconnaissance teams to speed reports of disasters' effects.

These and other reforms for the embattled Federal Emergency Management Agency (FEMA), announced [February 14, 2006,] by [the Department of] Homeland Security Secretary Michael Chertoff, are all well and good, say some experts. As the response to hurricane Katrina showed, the federal government's preparation for disasters has been so poor that there's much room for improvement.

But refurbishment of the bureaucracy alone may not address what a new inquiry called the root causes of the Katrina-response disaster—inattentiveness, incompetence, and lack of common sense.

"These changes may be all good things, but the question is how they work in practice," says Paul C. Light, a professor of public service at New York University.

Congressional Inquiries

After lying dormant for a few months, the issue of failures in the response to hurricanes Katrina and Rita have come roaring back with a vengeance in Washington.

Peter Grier, "Are FEMA Reforms Sweeping Enough?" *Christian Science Monitor*, February 14, 2006. Copyright © 2006 The Christian Science Publishing Society. All rights reserved. Reproduced by permission from Christian Science Monitor (www.csmonitor.com).

First, testifying before the Senate [in February 2006], former FEMA Director Michael Brown defended his reputation. He said the Department of Homeland Security, which includes FEMA among its parts, was interested only in planning to respond to terrorism, not natural disasters. Mr. Brown said he had informed the White House in a timely manner that New Orleans was flooding, and that he'd begged for food and water to be delivered to refugees crowded into the Superdome.

Refurbishment of the bureaucracy alone may not address . . . the root causes of the Katrina-response disaster— inattentiveness, incompetence, and lack of common sense.

Second, a Republican-dominated House [of Representatives] inquiry found that apathy, bad planning, and unheeded warnings contributed to the disaster, according to accounts leaked to news media. . . .

A draft of the House inquiry report said the federal government's response to Katrina was marked by "fecklessness, flailing, and organizational paralysis," according to the Associated Press.

Third, the Senate Homeland Security and Governmental Affairs Committee continued its inquiry into Katrina events, looking at alleged widespread abuse in federal emergency cash assistance programs for disaster victims.

Congressional investigators have found that as many as 900,000 of the 2.5 million applicants who received aid used duplicate or invalid Social Security numbers, or false addresses and names. "FEMA has a substantial challenge in balancing the need to get the money out quickly to those who are actually in need and . . . taking all possible steps to minimize fraud and abuse," stated an audit of FEMA programs, prepared by the Government Accountability Office and released by the Senate.

Chertoff Defends His Team

Faced with this storm of bad news, Secretary Chertoff responded with a forceful defense of his own and his department's competence. He rejected Brown's assertion that concern about terrorism trumped hurricanes, and he announced wide-ranging changes to FEMA, the first such steps taken since Katrina came ashore Aug. 29 [2005].

As many as 900,000 of the 2.5 million applicants who received aid used duplicate or invalid Social Security numbers, or false addresses and names.

In addition to the new employees, truck tracking, and recon teams, Chertoff's proposals include sending FEMA out to shelters to register victims for aid, rather than allowing them to register via phone and the Internet, and creating a database of pre-approved private contractors for debris removal.

Homeland Security officials will take "a hard, honest look at what we can do to improve our response capability," said Chertoff.

"Reform" vs. Common Sense

Given events such as the government's weak response to Katrina, "reform" is often the first watchword, as if perfection of process is always the answer. In some cases—such as widespread cases of fraud in FEMA aid—process was indeed the problem. "Even having a [New Orleans] phone book handy" would have improved the accuracy of FEMA employees handling aid requests, Mr. Light notes.

But general lack of focus can't be ordered up, or legislated. Process problems probably weren't the cause of the government's slow recognition of the unfolding disaster of the New Orleans flooding. Evidence of that was on the TV screen, for all to see; it just took too long for its implications to sink in at the White House and Department of Homeland Security headquarters.

"There's just no way you can order up greater common sense," says Light.

FEMA Is Not Ready for Another Katrina Disaster

Stewart M. Powell

Stewart M. Powell is a correspondent for Hearst Newspapers, publisher of six daily newspapers in the United States.

The Federal Emergency Management Agency [FEMA] is still not ready to respond to a Gulf Coast natural disaster akin to Hurricane Katrina despite its success in flooded Midwestern states.

That's the conclusion of an independent disaster researcher and key Texas officials who have monitored FEMA's far-flung response to flooding across Wisconsin, Iowa, Indiana, Illinois and Missouri over the past few weeks [of the summer of 2008].

An Agency in Flux

The hurricane season challenges facing the agency remain daunting, they said. Plans for evacuations and temporary housing must be made final, and residents must be persuaded to assemble emergency kits with a 72-hour supply of food, water and medicines to give first-responders time to reach them.

"It's hard to tell whether the FEMA response in the Midwest is a precursor for what FEMA can do and can't do," said Joseph E. Trainor, a staff researcher at the University of Delaware's Disaster Research Center. "The agency is in such flux even now, still trying to readjust from Katrina."

Floods No Comparison to Katrina

Hurricane Katrina had an apocalyptic impact in 2005. It claimed the lives of nearly 1,900 people, forced evacuation of more than 1.1 million others, ousted 770,000 from their homes and caused an estimated $96 billion in damage. Hurricane Rita followed quickly, claiming at least 120 lives and causing more than $11 billion in damage.

Authorities deployed 50,000 National Guard troops and more than 14,000 active duty military personnel.

The upheaval across the Midwest pales by comparison. Flooding across at least 3.4 million acres in the five states has claimed at least 22 lives, injured 149 and temporarily displaced thousands. So far, authorities have called in only 2,300 National Guard troops.

FEMA has received nearly 59,100 requests for assistance and doled out more than $124 million in housing and disaster-related assistance. The Small Business Administration has provided 226 emergency loans, valued at $14.3 million.

The limited demands on FEMA in the Midwest [floods of 2008] leave some . . . wondering whether the agency could respond effectively on a far larger scale.

Yet fewer than 500 flood victims sought refuge in Red Cross-style shelters at the start of the disaster, because so many neighbors stepped in to offer victims a place to stay.

"These are extremely low shelter figures for an event of this nature," said David Garrett, a top official in FEMA's disaster assistance operation.

The limited demands on FEMA in the Midwest leave some Texans wondering whether the agency could respond effectively on a far larger scale. Some 5,000 displaced families still live in travel trailers nearly three years after Katrina.

"The jury is still out," said Rep. Al Green, D-Houston, a member of the House Financial Service Committee that deals with housing issues.

"A work in progress," said Sen. John Cornyn, R-Texas.

FEMA would face challenges providing emergency housing and forcing evacuees to find temporary housing in communities across the country, said Michael Gerber, the top official of the Texas Housing and Community Affairs Department.

"It is very hard to plan for a mass migration of people—we saw that with Hurricane Rita," he said. "The big message here is that the state and federal government need to have done a stronger job being ready to meet that challenge, should it happen again."

Pilot Projects for Housing

FEMA's updated housing plan for the 2008 hurricane season emphasizes helping displaced victims initially find refuge in shelters. If victims' homes are severely damaged or destroyed, FEMA plans to move the displaced into nearby apartments, hotels and motels. Only as last resort will 3,500 mobile homes be called into action, many of the same FEMA trailers that stirred controversy after authorities found high levels of formaldehyde more than two years after Katrina and Rita victims began reporting illnesses.

FEMA would face challenges providing emergency housing and forcing evacuees to find temporary housing in communities across the country.

Looking ahead, FEMA is spending $400 million to assess different ways to provide that temporary housing with pilot projects in Texas, Alabama, Louisiana and Mississippi to test the use of modular homes capable of being transported by trailer.

The Harris County Housing Authority is awaiting final approval by FEMA and the Texas housing department to use part of a $16.5 million federal grant to create a 20-unit community of prefabricated homes manufactured by the Germany-based Heston Group, a firm that has provided housing for 90,000 soldiers in Iraq.

The $77,500 units, expected to be inhabited for no more than two years, are designed with 8-foot by 20-foot panels that can fit into a shipping container and be assembled by six workers in eight hours. The first of the 20 units has yet to be erected in Houston. Another 130 units are expected to be erected in other East Texas areas affected by the 2005 hurricanes.

FEMA Officials Confident

FEMA officials remain upbeat about their ability to win back public confidence.

"The response to the floods in the Midwest demonstrates clearly that the FEMA of today is not the FEMA of 2005," said Glenn Cannon, an agency official. "We have risen to the level of what the American public expects from us."

Homeland Security Secretary Michael Chertoff said FEMA, an agency in his department, has improved tracking supplies, established advance contracts for bus evacuations, and added more full-time disaster response specialists.

"We still have more work to do," Chertoff said. "I think certainly in the response area, all the instances we've had over the last year, FEMA is much quicker, more responsive. . . . It's generally a more capable agency."

FEMA Failed to Develop a Housing Plan for Hurricane Ike Evacuees

Dahleen Glanton

Dahleen Glanton is the southern United States correspondent for the Chicago Tribune.

In its first major test in three years, the Federal Emergency Management Agency [FEMA] has come under scrutiny for failing to develop a long-term housing plan for the more than 1 million evacuees from the Texas Gulf Coast in the aftermath of Hurricane Ike.

Faced with criticism, FEMA [in September 2008] agreed to pay a month of hotel expenses for some evacuees from the hardest-hit areas. But in a meeting with Homeland Security Secretary Michael Chertoff on [September 19, 2008], local officials expressed concern that there was no longer-range plan for residents whose homes in devastated areas such as Galveston, Beaumont, Port Arthur and Orange will be uninhabitable indefinitely.

"Housing is a big concern, and our county judges and elected officials brought it up with Secretary Chertoff," said Officer Crystal Holmes, spokeswoman for Beaumont Emergency Operations. "We just don't have apartments here, and when you think of the thousands of homes actually obliterated, where are all those people going to live?"

Housing Strain

With 32,000 people in shelters across the state and thousands more living in hotels and with relatives or friends, Texas officials said they are anticipating a housing strain on the area, which already has a shortage of apartments and other rental units. Meanwhile, the housing burden has fallen on state shelters, which were set up as an emergency resource and now could be forced to remain open longer.

According to Zachary Thompson, director of the Dallas County Department of Health and Human Resources, FEMA should have established programs with housing agencies across the state before the storm hit or immediately afterward so that apartments and government-subsidized housing could be readily identified. Those are lessons that should have been learned from Hurricanes Rita and Katrina in 2005, he said.

"We have seen this movie before. It happened with Katrina," Thompson said. "When you evacuate the majority of residents from an impacted city, the game plan for the federal government should be to look at housing needs. People clearly can't go back to Galveston. The shelters were put in place to get people out of harm's way. The next step is up to FEMA. No city in America is set up to handle long-term shelters," he said.

Thompson said Dallas was consolidating 4,000 evacuees into one location at the Dallas Convention Center. In Ft. Worth, officials said they would ask churches to open extended shelters, and in San Antonio, authorities struggled to find clothes for evacuees, some of whom had worn the same items for five days.

We have seen this movie before. It happened with Katrina.

FEMA spokesman Marty Bahamonde said the agency is working with real estate agents in Houston and other parts of Texas and Louisiana to identify vacancies. He said the agency would pay rentals for up to 18 months. "We try to find temporary housing and hope things will change down there so they can go back in 30 days," said Bahamonde.

He said FEMA also is looking for mobile homes in the area that can be rented, but the agency has abandoned the controversial travel trailer program, which provided housing for Katrina and Rita evacuees for months after the storms. FEMA was criticized for taking too long to get people into the trailers initially and for allowing people to stay in them after learning the units were a health risk.

Local officials have urged residents not to return to the flood-ravaged areas until basic services such as electricity, sewer and water are restored. Still, thousands of people have tried to return to Galveston and other areas, despite a shortage of food, water and ice.

Supplies Sit Idle

Meanwhile, about 1,000 18-wheelers filled with food, ice, water, cots and tarps have sat idle at a staging area on Interstate Highway 10 on the edge of Beaumont for days, while FEMA and the state blamed each other for failing to get the supplies to residents.

FEMA and the state blamed each other for failing to get the supplies to residents.

"We don't care whose fault it is, we just want the assets to roll," said Holmes, the spokeswoman for Beaumont Emergency Operations.

Economics is the main reason many residents do not evacuate, officials said, and one of the reasons they are eager to return. With gas prices hovering at $5 a gallon, eating out and hotel rooms at $150 a night or more, it could easily cost $1,000 for a family to evacuate.

Orange County Judge Carl Thibodeaux, who is responsible for managing the coastal county, said the federal government should pay for gasoline, food and housing costs when residents leave under mandatory evacuation orders.

"One issue is the enormous expense of evacuating. I had to tell people to leave for [Hurricane] Gustav, and when the storm was over, there was nobody there to help them. They didn't qualify for any aid from the federal government because there was minimal damage," Thibodeaux said. "Ten days later, I had to ask them to leave again for Ike."

According to FEMA officials, when a city or state orders a mandatory evacuation, it does not mean the federal government will pay for it. "We simply encourage people to evacuate for their own safety, and sometimes there is a cost associated with that," Bahamonde said. "We understand that it is a hardship, but we have seen too many times what happens when people don't evacuate. They die."

To qualify for hotel reimbursements, residents must prove their homes are uninhabitable, which does not include the loss of electricity. Most residents in Galveston, Beaumont, Port Arthur and Orange would be included, but not Houston.

Officials said 4,000 people had checked into hotels under the program and another 107,000 have qualified for it. However, there aren't enough hotel rooms. Many hotels in the area are closed and others are booked through October with emergency workers and evacuees who got in early.

CHAPTER 3

Should Citizens Rely on the Government to Respond to Disasters?

Chapter Preface

The effectiveness of any government disaster response is dependent on a fundamental step in the disaster response plan—persuading citizens to evacuate their homes in the face of an impending disaster. Recent disasters have shown, however, that even when public officials issue mandatory evacuation orders and give dire warnings about the risk of not evacuating, a fair number of residents typically refuse to leave. They take this risk for many reasons, but the reasons are often related to the actual or perceived ability of the government to protect people, pets, and property during times of crisis.

At the time of Hurricane Katrina, for example, more than one hundred thousand residents of New Orleans stayed in their homes, disregarding Mayor Ray Nagin's mandatory evacuation order and his dire warnings about the size and danger of the approaching storm. After the hurricane hit, many of these people were found on their roofs with little food or water, hoping for helicopter rescue after storm waters rose up through their attics. Some people are believed to have died when the flood waters rose higher than their rooftops. Some experts explained later that many of these people simply did not have the financial ability to evacuate, either because they lacked transportation or did not have enough money to pay for gas, food, and a hotel for what promised to be an indefinite period following the storm. Critics argued that the federal, state, and local response plan during Katrina should have included a better strategy for moving poor people who did not have the means to leave on their own.

Some people stayed in their homes during Hurricane Katrina because the shelters and evacuation centers did not allow pets or other animals. Given the choice between staying home or leaving their pets behind, they stayed in their homes, refusing to be separated from their beloved pets. This loyalty to pets should not be surprising, experts say. Studies by the American Veterinary Medical Association and other animal-focused organizations show that between 60 and 70 percent of American households have pets, and the majority of these consider the pets to be members of the family. A 2007 survey by the American Humane Association found that 47 percent of Americans would refuse rescue assistance if it meant evacuating without their family pets. This information highlights the importance of rescuing both animals and people during disasters. This conclusion was reached by congressional investigators following the Katrina debacle, resulting in the enactment of the Pets Evacuation and Transportation Standards Act (PETS), which now requires the inclusion of companion animals in disaster planning at the state and local levels.

A survey released in 2008 by Columbia University's National Center for Disaster Preparedness found that Americans' attitudes toward protecting their children are another significant factor in deciding whether or when to evacuate. The survey found that in the event of a disaster, 63 percent of parents would defy evacuation orders and instead try to pick up their children at school or daycare. During sudden disasters where there is little or no advance warning, such as a nuclear or terrorist disaster, experts say such a response could result in traffic jams or pose other problems for government disaster relief, possibly causing needless death or injury to many citizens.

Another reason that people sometimes choose to stay rather than obey evacuation orders is that they underestimate the dangers posed by the impending disaster. People who have lived along the Gulf Coast a long time have ridden out many hurricanes in the past; in the case of Hurricanes Katrina, Rita, Gustav, and Ike, they may simply have bet that they could ride out another hurricane as well. Residents also may have heard so many warnings from public officials over the years about the dangers of storms that did not materialize, or were less serious than announced, that they assumed the storm would not be as bad as predicted—a case of authorities "crying wolf" too many times. They stayed because they had grown weary of disrupting their lives and leaving their homes. Experts have speculated that some people, too, may not understand disaster warnings because they are ill, hearing-impaired, isolated in their homes without communication from the outside world, or do not speak English.

In the case of certain types of disasters, some home and business owners decide to stay behind to protect their property, either from looters or from the disaster itself. This phenomenon is often seen in the western United States, where massive wildfires frequently destroy large swaths of acreage, sometimes in residential areas. The reason given for staying in the path of a dangerous fire is usually that homeowners do not believe that the government will be able to protect their property. As Kirk Gafill, general manager of the Nepenthe Restaurant in Big Sur, California, put it after he and five employees stayed up all night trying to protect his business during the 2008 wildfires that hit the region, "We know fire officials don't have the manpower to secure our properties. . . . Based on what we saw during Katrina and other disasters, we know we can only rely on ourselves and our neighbors."

Authorities are reluctant to handcuff or forcibly drag people from their homes, although some commentators have suggested that this might be legally possible in cases where residents are endangering themselves or others by staying. More often, people who refuse to leave during evacuation orders are warned they may not be rescued and are asked to supply information on next of kin and dental records, so that if they die, they can be identified.

The viewpoints in this chapter set forth some of the differing opinions on whether citizens can or should rely on the government in the event of a natural disaster or other emergency.

FEMA Is Ready to Respond to Major Disasters

John Solomon

John Solomon blogs about emergency management issues on his Web site, In Case of Emergency, Read Blog, *and is writing a book called* In Case Of Emergency, Read Book: Simple Steps to Prepare You and Your Family for Terrorism, Natural Disasters and Other 21st Century Crises.

In an interview [in August 2008] on the eve of the third anniversary of Hurricane Katrina, FEMA [Federal Emergency Management Agency] Administrator R. David Paulison told me that the agency has been "rebuilt" and is "ready" for the next major hurricane or other disaster. Unfortunately, he may have a chance to test his assessment in the next several days as Tropical Storm Gustav heads toward the U.S. mainland threatening the Gulf Coast with what forecasters say could become a Category 3 hurricane. Paulison is now in the region as preparations are being made for a possible Monday [September 1, 2008] landfall.

A Better FEMA

Paulison says he wants Americans to know that FEMA has worked very hard to overhaul itself in the years since Katrina. "This is a responsive organization again," he says, adding, "There has been a change of culture. We have gone from a reactive to a proactive agency."

Paulison took over FEMA in April, 2006, and the agency has received largely good reviews for its work in recent major disasters including this summer's floods in the Midwest and last year's Southern California fires. The agency has doubled its budget and distributed $25 billion in preparedness grants to communities around the U.S, according to Paulison. "The public should know that FEMA is ready."

FEMA has worked very hard to overhaul itself in the years since Katrina.

Citizens Must Also Be Prepared

However, he emphasizes that FEMA and local authorities are relying on the public to do its share when it comes to emergency preparedness. In fact, Paulison says that citizens are an integral part of the nation's disaster response effort by taking care of themselves and their families (as well as neighbors who need assistance) and therefore lightening the load on emergency personnel. "All the states, local communities working together cannot take the place of personal responsibility for taking care of yourself," notes Paulison.

He says every citizen should have emergency supplies, a communications plan and know where they would go in the event of an evacuation. "If individuals don't take that responsibility, this country is not going to be ready and be able to take care of everyone." Paulison says emergency personnel rely on those who can help themselves to do so so they can focus on those who, for physical, mental, logistical, or financial reasons, cannot.

Yet, Paulison acknowledges the difficulty of getting the personal readiness message through to much of the public. "We need to return to a culture of preparedness. We've gotten away from the preparedness ethic of the 50's and 60's."

To me, a way to engage the public and spur a greater commitment to preparedness might be to more explicitly communicate the two choices (one positive, one negative), as described by Paulison, that citizens can make when it comes to personal emergency preparedness: you can either be part of the problem and make things more difficult for first responders, but you also can be part of the solution by helping them by what you do before and during a disaster.

FEMA and local authorities are relying on the public to do its share when it comes to emergency preparedness

---

This public messaging approach would combine a bit of both the stick and the carrot. It would also underscore to the citizenry that emergency responders see them as an integral part of the nation's disaster preparedness and response effort, something I do not think that most people realize. It would also help explain the citizen's role and responsibility in emergency preparedness, and give them a better sense of where they fit in the bigger picture.

I don't believe the government should be afraid to explicitly tell the public that each of us can either hinder or help relief efforts by what they decide to do before and during a disaster. And, that it's up to each of us to choose. I think that's a choice and a challenge that might get people's attention and maybe lead to action.

The Private Sector Cannot Take the Place of Government in Disaster Response

Dante Chinni

Dante Chinni is a journalist who writes a regular media column for The Christian Science Monitor, *an international newspaper.*

It didn't really take that long, not even a month. After careful consideration some of the leading conservative OP-ED [opinion-editorial] minds in the country have pinned down the problems with the federal government's response to Hurricane Katrina. The problem was ... the government itself.

Disaster Response from Wal-Mart

The argument goes something like this: The folks in this town [Washington, D.C.]—with all their red tape, their bureaucratic mumbo jumbo, their chronic inability to get anything done— simply can't be expected to handle emergency management. It would be better to turn it over to the private sector, or, as *The New York Times*'s John Tierney suggested, to Wal-Mart.

Wal-Mart, you see, responded quickly to the disaster on the Gulf Coast. While the government was busy twiddling its thumbs, the nation's largest retailer was sending truckloads of ice and generators and chain saws to the needy in the region.

Mr. Tierney, speaking on [conservative journalist and news analyst] Tucker Carlson's *MSNBC* show, told the audience, "[Wal-Mart's] got its own emergency operation center where, I think six days before the hurricane, as soon as that storm appeared, they started moving generators and supplies and trucks into position and they're ready. I mean this is what they do all the time and they, you know, they're efficient."

Dante Chinni, "Wall Street Shouldn't Trump the Government in Emergency Response," *Christian Science Monitor*, September 27, 2008. Reproduced by permission.

Indeed they are. The question is would we now simply say that we don't believe the government can be efficient in times of crisis? It has in the past. And since when is rushing materials to an area that needs it suddenly the province of geniuses— something only those whizzes on Wall Street could have come up with?

Wal-Mart . . . responded quickly to the [Katrina] disaster on the Gulf Coast.

Blaming Government Is Too Simplistic

Just maybe, the government's real problem in responding to Katrina wasn't the fact that it was swimming in a sea of public sector inefficiency, but the fact that in this particular case these particular agencies were blatantly incompetent. Maybe some people in the Homeland Security Department and the people running FEMA [Federal Emergency Management Agency] and, yes, some local officials down in the Gulf made some big mistakes.

The blunders that followed Katrina may be a very nasty indictment of the [George W.] Bush administration, or of Louisiana governor Kathleen Blanco or of New Orleans mayor Ray Nagin or all of the above. But if we are going to see those errors as an indictment of government in general, we need to apply the "errors equal failure" standard across the board, to business as well as government.

For instance, should we allow any private companies to supply electric power now that we've witnessed [bankrupt energy company] Enron's seedy activities? How can we keep giving contracts to private companies when we find out later they can't account for billions of dollars? And you can pick from a long list of culprits on that one, though Halliburton [U.S. company contracted to oversee oilfields in Iraq] springs to mind.

None of which is to say the private sector is always bad or inefficient, it's just that the "government is the problem" approach is too simplistic. It assumes that efforts succeed or fail because of motive. That is to say, businesses succeed because money is at stake and government fails because there is no accountability or profit motive.

The government's real problem in responding to Katrina ... [was] that in this particular case [its] agencies were blatantly incompetent.

Beyond the fact that this formulation simply ignores some realities (the US military's ties to the dreaded public sector don't seem to get in the way when it's needed), there are two big problems with this thinking.

First, there are without question many successful and soundly built private enterprises in the United States producing excellent goods and services, but there are also many, lousy, inefficient companies that produce garbage. In other words, using airlines as an example, for every "brilliant" Southwest, there's a "suffering" Northwest. The differences between the two isn't the profit motive, it's the minds at the top and the people inside.

Second, and more important, to say there's no accountability in government is simply wrong—and if you watched the way the White House responded to the arrival of Hurricane Rita you know they understand that.

If you didn't know any better you might have thought President Bush had appointed himself head of FEMA. Here he was at a FEMA meeting in Washington. There he was in the Rocky Mountains' NorthCom Headquarters monitoring the effort. It may have all been a bit showy, but the message was unmistakable. The president had read his poll numbers and knew he was being held responsible for FEMA's gaffes after

Katrina. He was determined to do anything and everything to run a better show this time around.

It wasn't just the presidential photo ops, though. Federal, state, and local officials went out of their way to sound the alarm and to be prepared. Those who didn't have transportation in Galveston were given bus rides. The military was on the ground before the storm hit.

And they did it all without Wal-Mart.

Citizens Cannot Rely on FEMA During Catastrophic Events

Dorothy A. Seese

Dorothy A. Seese, a Christian conservative and former child actress, has a degree in political science.

There are millions of Americans not in the disaster area of the Gulf states who are noticing that the surrounding communities seem to be the first on the job. The one exception is the Coast Guard, the most underrated and overlooked branch of the government, now part of the Department of Homeland Security. It isn't the fault of the military that they don't respond in a timely fashion, the blame lies directly on the shoulders of a mismanaged, bulky, tossed-together Federal Emergency Management Administration [FEMA]. FEMA is its own disaster. It's a good thing neighboring cities and towns volunteer their own people to work our disaster areas. And it's pathetic that FEMA has so much power when it can do so little in a real disaster.

Do Not Rely on FEMA

If FEMA is all you rely on for aid in a catastrophic event, then you're in trouble. The town up the road is a better source of help.

Of course, disaster management by voluntary cooperation, minus bureaucratic red tape, has always worked better and faster than what the federal government provides. As one television reporter put it, "we are our brother's keeper." What has come out of hours of watching news and weather reporting is my impression that while far too many people elected to stay in a hurricane-prone area after being ordered to evacuate, it is the communities and their mayors that shoulder the burdens of first response. Their resources may be limited but something at least gets moving, some folks live who otherwise would die. They use what they have without waiting for some piece of paper.

If FEMA is all you rely on for aid in a catastrophic event, then you're in trouble. The town up the road is a better source of help.

There is no excuse in America for people to die because of floodwaters, lack of electricity or phone service, or even absence of safe roads in and out of the area. We have the equipment, we can't get it mobilized. The locals cannot breach the airbases nearby and grab the choppers they need or the supplies.

If the New Orleans disaster proved one thing, it is that American government, four years past Nine Eleven [i.e., September 11, 2001], still is not prepared to do a turn-on-a-dime emergency response. FEMA may have a role to play, largely in handing out checks at the proper time and hopefully to the proper authorities to pay for the services of those who earned their pay and those who lost all their resources. As a first responder, it appears to be useless.

Well—no one expected Hurricane Katrina to be so devastating. My reply is, are emergencies supposed to be timed like a movie script or confined to certain magnitudes of disaster? Oh go away! The federal government's National Oceanic and Atmospheric Administration (NOAA) knew two days prior to the landfalls near Grand Isle, LA and then Gulfport, MS, that they had an abnormally large hurricane on the move. I'm a weather watcher and that monster took up over half the Gulf of Mexico tip to tip! I tried to explain it to a friend who called. She wasn't impressed. Later that evening she called me gasping in disbelief at the size of Katrina. I tried to tell her, she had not grasped it until she saw it. But what I saw, and she saw, was seen by millions of Americans, even the twits who didn't have enough brains to leave the area when ordered to do so. Even by the government that brags on the efficiency and usefulness of FEMA. Yes, there were those who could not leave, they didn't have transportation. Why? Couldn't someone have commandeered all the school buses in the area? So far, everything the mayor of New Orleans has said sounds low key and slow motion. He was safe so what the hell? It's a shame every city has its village idiots but it's a sorry state of affairs when the public officials are among them!

The Idiocy of Building near the Gulf

My bias against the Gulf Coast is no secret. The Seese line to which I belong came to Baton Rouge some time between 1830 and 1845. But at least they stopped at Baton Rouge and stayed away from the soup bowl called New Orleans. If they had not, it is doubtful if any of the present-day Seeses would be here, as the flooding of New Orleans is not a new thing. The great Labor Day hurricane of 1935 occurred when I was not yet two months old, but my parents and I were living close to Santa Monica, California. I've been to New Orleans twice, once in 1987 as a one-day visitor staying in Gonzales, LA with friends. The second time was in March of 2004 to visit my cousins who had found me via the internet. The cousin who found me is a retired surgeon. He had a huge home in a suburb of New Orleans and a two-story beach house in Pass Christian [Mississippi] that he finished rebuilding eight months ago from a previous hurricane's devastation. Pass Christian has been described on the news reports as "leveled." I don't know about the home in Jefferson Parish [i.e., the New Orleans suburban house]. It may or may not be standing.

This rebuilding after hurricane disaster and destruction is beyond all reason. Yes, some years the Gulf Coast doesn't get hit by hurricanes. But it has and it will again. When disaster lurks with every summer hurricane season, it seems a bit unreasonable to even issue building permits for beach homes or allow commercial construction near the Gulf. Their economy is supposedly built on tourism and gambling, sinking sand enterprises that can come and go with the economic tides. Apparently the economies of these states are as vulnerable to disasters as their topography. While I admire their spirit, it seems terribly misdirected.

In northern Arizona, we have a gully called the Grand Canyon and it is a national park. No condos hang precipitously on the sides of the canyon walls. It's a national park. People go there to gawk and take photographs. Some go to paint. Others just go there because it's there. If building were allowed there, the developers would descend on it like vultures on a rabbit carcass.

The Gulf Coast seems like an ideal place to establish a national park where tourists could go gawking, but not building at or below sea level. Engineers and meteorologists could establish the average surge devastation inland, and no building should be allowed any closer to the Gulf than that perimeter. I believe in freedom, but I also believe in traffic lights to control wild drivers.

And I believe the situation in New Orleans is proof enough that our federal government has a nice followup program that is supposed to be among the first responders—it costs enough.

When disaster lurks with every summer hurricane season, it seems a bit unreasonable to even . . . allow commercial construction near the Gulf.

It's also feasible to believe that in spite of other impending disasters, New Orleans will be rebuilt on sinking sand with bigger pumps and a few other technological improvements rather than rebuilt as a theme park city where tourists can go see replicas of what was once "New Orleans."

I'll go on record as being against N'Awlins, against Mardi Gras, and against allowing any Gulf shore property to be anything more than a national park and tourist destination. Put the hotels, the homes and the businesses inland.

You know they won't do that. They'll rebuild, and rebuild, and rebuild with each devastating blow from the forces of things greater than man's ability to manage. Now that's stupid!

Most Texans in Hurricane Country Do Not Rely on the Government

Gina McCauley

Gina McCauley is an attorney and founder of the blog WhatAboutOurDaughters.com.

As with every storm, Hurricane Ike [in 2008] will be either a boom or bust for politicians. Some will reaffirm the trust that voters placed in them. Others will embarrass their constituents on a national stage. Some politicians shine in the spotlight. Others just crash and burn. Bill White, the mayor of Houston, Texas, and Ed Emmit, a Harris County judge, have shone, going so far as to camp out and become dispatchers when relief supplies bottlenecked, leaving people waiting in line for hours for ice that was sitting in a parking lot at a sports arena.

To their south however, Lyda Ann Thomas, the mayor of Galveston, appears to be going through an emotional meltdown. On [September 16, 2008], Thomas announced that Galveston residents would be allowed back on Galveston Island briefly. Predictably to everyone but Thomas and her city manager, thousands rushed back to the island, clogging the major highway with a 10-mile long traffic jam and burning already scarce gasoline. Thomas quickly reversed her decision, leaving thousands of her constituents parked on the highway waiting to enter, justifiably irate.

The next day she was reduced to tears during a meeting of the remnants of the Galveston city council when a council member accused Thomas of showing favouritism by allowing a selected few to return to the island and providing essential services such as tetanus shots to city employees while denying them to residents that remained on the island.

The Federal Emergency Management Agency (FEMA) remains the whipping boy for everything that goes wrong in the aftermath of the storm.

There are some things that have changed about hurricanes in recent years. Hurricane Katrina hangs over the head of every coastal area politician. Nobody wants to be the next Ray Nagin (the mayor of New Orleans), Kathleen Blanco (the former governor of Louisiana) or [President] George [W.] Bush. Even Bush doesn't want to be Bush. He took to the airwaves to send a message that he was aware that a hurricane had struck the fourth largest city in the country.

For his efforts, the Federal Emergency Management Agency (FEMA) remains the whipping boy for everything that goes wrong in the aftermath of the storm. "Blame FEMA" is the new post-hurricane mantra. I didn't even know FEMA existed as a child. Prior to Katrina, most Americans likely didn't know what FEMA did, and apparently some local officials still don't know.

Relying on Family and Friends

Despite news reports to the contrary, most Texans aren't relying on FEMA. They are relying instead on what we have always relied on in hurricane country: the FFEMA, the Family and Friends Emergency Management Agency.

Our family disaster response swung into action when I got the call from my mother at 7:00am a week ago. Jefferson County officials had ordered a mandatory evacuation of my hometown in southeast Texas. My parents and extended family hadn't bothered to unpack from their evacuation for Hurricane Gustav a little over a week earlier. My mother wanted to head west toward me, but ended up heading north.

After that initial call, as is typical, the cell phone circuits were jammed and calls could not go through. Text messages became the only reliable means of communication. My parents didn't know how to text message, however. My sister on the east coast sent me a text the following morning that, after travelling overnight, my parents and assorted extended relatives had finally reached Arlington, Texas, where they were near family.

Most Texans aren't relying on FEMA. They are relying instead on ... the FFEMA, the Family and Friends Emergency Management Agency.

My two sisters who live in Houston messaged that they had decided to ride the storm out. My sister in the suburbs north of Houston lost an oak tree, but they survived Ike with no major structural damage. After a day and a half without electricity however, the novelty of playing board games in the dark got old, and they messaged that they were coming to stay with me in Austin.

Like my family, most people aren't waiting around for the federal or state government. They are clearing their streets, creating their own rudimentary power grids by draping extension cords across streets and driveways to share electricity with neighbours. Family members in unaffected areas are driving into town with coolers full of ice and five-gallon gasoline canisters strapped to roofs of cars and SUVs. It was a bit disconcerting driving on I-35 watching an SUV with six full five-gallon gas canisters strapped on top. Is that safe?

People are doing what they have always done after major storms; they are adapting to the situation. Which is why I taught my 60-something-year-old mother how to text message on Sunday when I drove to check on them at their hotel in north Texas. Cell phone companies can provide GPS, internet service and picture mail, but they haven't quite figured out how to cope with all of the families and friends who clamour for contact with each other in times of disaster.

People are doing what they have always done after major storms; they are adapting to the situation.

Mama figured it out in about three minutes. In her exuberance in sending and receiving messages I had to caution her that text messaging is not part of her cellular plan. Her newly discovered modern day Morse code was probably costing her $0.15 a message. She didn't appear to care. Contact with the outside world was more than worth it.

Reliance on the Federal Government Gives a False Sense of Security to State and Local Governments

Matt A. Mayer, Richard Weitz, and Diem Nguyen

Matt A. Mayer is a visiting fellow at the Heritage Foundation, a conservative policy institute, and an adjunct professor at Ohio State University. Richard Weitz is a senior fellow and director of program management at the conservative Hudson Institute. Diem Nguyen is a research assistant at the Heritage Foundation.

The increasing tendency since 9/11 to look to Washington for every answer regarding disaster response is troubling. The insistence that the Federal Emergency Management Agency (FEMA) play an ever-expanding role in addressing day-to-day emergency responses is hindering, not strengthening, the agency's ability to prepare for the next national catastrophic disaster. Even worse, as the federal government pledges to improve its response, state and local governments are getting a false sense of security, relying on Washington rather than preparing proper emergency responses themselves.

The October 2007 wildfires in California provide a revealing glimpse into the continued federalization of disasters. Trumpeted as proof that Washington is ready for the next Hurricane Katrina, California's response really demonstrates that well-organized state and local efforts are far more critical than federal ones. Rather than encourage more Washington-centric solutions, Congress and the White House should focus on lessening the federal role in day-to-day state-level emergencies and emphasize a greater responsibility among state and local communities for preparing and developing response plans for local disasters.

Federalization of Disasters Continues

This year alone [2008], FEMA has issued 35 disaster declarations: 15 major disaster declarations, three emergency declarations, and 17 fire management assistance declarations. FEMA is on pace to issue about 144 disaster declarations in 2008, which would be the third-highest number of disaster declarations since 1953.

The increasing tendency since 9/11 to look to Washington for every answer regarding disaster response is troubling.

The record of 157 declarations achieved in the 1996 election year under President Bill Clinton's FEMA Director James Lee Witt still stands despite the best efforts of all levels of government to get Washington to foot the bill for as many disaster responses as possible. Even more troubling, President George W. Bush's yearly average of disaster declarations will hit 130 by the end of his Administration—an almost 50 percent increase over President Clinton's yearly average. . . .

These trends are bad for emergency management and bad for federalism. As the federal government participates more in disaster response, states will rely more heavily on that federal presence and, as an inevitable result, will be less prepared and less equipped to deal with both contained calamities and truly catastrophic events like Hurricane Katrina. The key to successful emergency management is a quick and effective response from state and local communities, which can react in a timely manner and are much more prepared and trained for the particular disasters that often occur in their specific regions.

New FEMA Fuels the Federalization Fire

Following Hurricane Katrina, the federal government came to new conclusions about how to improve disaster response.

FEMA stated that part of the reason that the wildfire response was such an effective "team effort" was that people now "don't wait to be asked" to offer help in a crisis. The Administration seems to have learned from its slow response to Hurricane Katrina and "doesn't want to be bitten again," according to former FEMA Director Joe Allbaugh. For example, President Bush called California Governor Arnold Schwarzenegger to offer help before Schwarzenegger had even asked for federal assistance. President Bush subsequently anticipated the state's request to declare a state of emergency and approved it just one hour after the request had been filed.

The key to successful emergency management is a quick and effective response from state and local communities.

Pentagon officials say that Hurricane Katrina taught them to be more "forward leaning" as well. In the words of Paul McHale, Assistant Secretary for Homeland Defense, "One of the lessons that we, as a nation, learned is that in a crisis, you don't wait to be asked; you lean forward, you prepare your capabilities and you ask, very pointedly, 'How can I help?'" Even before California authorities requested help, the National Guard Bureau deployed military aircraft to California on a training mission, placing them in a better position to help fight the fires.

In addition to offering federal assistance to state and local governments, FEMA has provided $4,571,714 to rebuild homes destroyed by the fires. FEMA also has provided California with a substantial amount in homeland security grants to equip the state with fire-fighting equipment: Between 2001 and 2007, California fire departments received $147 million under the Assistance to Firefighters Grant Program.

The "forward leaning" approach may have been beneficial for California, but it will leave the nation less prepared to deal with catastrophes the size of Hurricane Katrina. It requires a greater increase in responsibility for the already overwhelmed FEMA, leaving the agency even more thinly spread and ill-prepared for a catastrophe affecting thousands of American lives. An over-eager federal government also creates a false mindset in state and local governments—the expectation that they can rely on the federal government for help—that will leave them less prepared to respond effectively in the critical first 72 hours.

With this federalization of disaster management, FEMA has bitten off more than it can chew. As noted above, between 1993 and 2007, FEMA tripled the number of declarations issued each year. As a result, it is responding to a new declaration every three days. As more resources are devoted to this increased response, local preparedness withers on the vine. In contrast to the increase in declarations, FEMA's budget and employees have not grown by proportional amounts. In 2006, FEMA had 2,000 employees, 500 *fewer* than in 1992. In addition to a smaller workforce, the budget increase was barely perceptible—from $4,834,065,000 to $4,834,744,000.

Given these trends and resource limitations, FEMA, despite its best efforts, will likely prove inadequately prepared for the next catastrophe. . . .

State and Local Responses Were Key

In general, the response to the California wildfires was mostly successful and a marked improvement from the response to the similar fires of October 2003. The most notable characteristic of the 2007 response, and a pivotal factor in its success, was the proactive nature of the state and local responses. Unlike after Hurricane Katrina, the response to the California wildfires was state and locally driven, not federally driven. State and local leaders made a vigorous effort to take charge and avoid visible infighting. Governor Schwarzenegger took charge of the situation early instead of waiting for federal officials to address the problem.

California's governor flew by helicopter to each firefighting base, meeting with local officials and passing on their requests to the federal government, and followed up by returning frequently to these locations to verify that the assistance had arrived. The state's ability to assess quickly what type of assistance was needed enhanced its ability to work effectively with the federal government. Local leadership and initiative were particularly important considering that San Diego lacks an integrated fire department and relies on a "hodgepodge of local departments that are almost all serving areas where populations are growing faster than their tax bases."

With this federalization of disaster management, FEMA has bitten off more than it can chew.

California's take-charge stance toward firefighting was not the only factor in its successful response. Compared with other states and regions, southern California is a well-prepared area that possesses a formidable emergency response team.

In the October 2003 wildfires, the response was unorganized and chaotic. Local responders absorbed the many lessons learned from that disaster and applied them to the 2007 wildfires. Because of communication problems in 2003, 911 reverse calling, where operators contacted households advising them to evacuate, was used for the first time in 2007, and the system worked well.

The federal government did play a role in the response, though it was not critical. Bush Administration officials, mindful of the criticisms they had received after Hurricane Katrina, adopted a highly visible and proactive stance throughout the wildfire crisis.

After Katrina, FEMA made an important change in how quickly it engages in disasters. FEMA used to wait for a disaster to overwhelm state and local officials before it intervened. To ensure the timely arrival of assistance, FEMA now begins moving some hard-to-deploy assets into an affected area even before local authorities request federal assistance. This approach works for a hurricane or wildfire, where most people have advance notice of the approaching danger, but may not be too useful for an earthquake or terrorist attack, near-instant disasters that usually catch people off-guard.

Finally, past U.S. Department of Homeland Security (DHS) federal grants for first responders proved vital during the response to facilitate coordination and communication among city, county, state, and federal government personnel. DHS has provided $1 billion to help states and cities improve communications interoperability. This spending paid off on the ground during the 2007 wildfires as many of the first responders commented on the smoothness of communications and interoperability.

The Los Angeles County fire chief told *The Wall Street Journal* that good communication with other state and federal agencies had led to improved coordination among firefighters. The improvements in interoperability led FEMA Director David Paulison to remark, "What we see now that we did not see during Hurricane Katrina is a very good team effort from the local, the state, and the federal government and across the federal agencies."

If there was one important lesson to take away from the wildfires, it is that well-prepared state and local governments are crucial to the efficacy of a response. Because state and local governments are always the first to respond to a disaster, they must be specifically prepared and should not depend on the federal government. As former Florida Governor Jeb Bush wrote a few weeks after Hurricane Katrina:

> As the governor of a state that has been hit by seven hurricanes and two tropical storms in the past 13 months, I can say with certainty that federalizing emergency response to catastrophic events would be a disaster as bad as Hurricane Katrina.

Just as all politics are local, so are all disasters. The most effective response is one that starts at the local level and grows with the support of surrounding communities, the state and then the federal government. The bottom-up approach yields the best and quickest results—saving lives, protecting property and getting life back to normal as soon as possible. Furthermore, when local and state governments understand and follow emergency plans appropriately, less taxpayer money is needed from the federal government for relief. . . . If the federal government removes control of preparation, relief and recovery from cities and states, those cities and states will lose the interest, innovation and zeal for emergency response that has made Florida's response system better than it was 10 years ago. . . . But for this federalist system to work, all must understand, accept and be willing to fulfill their responsibilities.

Yet, as noted above, over the past 16 years, Washington has federalized even the most routine of disasters, which means that the federal government pays the bill. . . .

It is time to get back to our federalist tradition and empower state and local governments to take the lead in managing disasters.

Federalization Not the Solution

As the California wildfires and the latest FEMA data show, the federalization of disaster responses continues to accelerate. This practice must end. It stretches FEMA's already strained resources even thinner and encourages state and local governments to divert their disaster-response resources to more immediate needs like transportation, education, or health care.

While it is understandable that members of FEMA and the Bush Administration want us to believe that they have fixed what was so horribly broken, the politicians in those entities need to temper their public relations machines or risk deepening the false sense of security and apathy that already exists to a large degree across America. The risk of catastrophic terrorist attacks is real, and it is high. Much work remains to be done to prepare for these probabilities. It is time to get back to our federalist tradition and empower state and local governments to take the lead in managing disasters.

CHAPTER 4

How Can U.S. Disaster Response Be Improved?

Chapter Preface

At the time Hurricane Katrina hit the United States in 2005, the response of the Federal Emergency Management Agency (FEMA) was guided by the National Response Plan—a plan of action that set fixed rules for how FEMA should respond to a disaster. This plan emphasized terrorist emergencies and, among other mandates, required the Department of Homeland Security (FEMA's parent agency) to declare a disaster an incident of national significance before initiating a federal response. In the case of Katrina, getting this declaration took days, and many critics said this delayed FEMA's ability to act.

On January 22, 2008, Homeland Security secretary Michael Chertoff announced the "National Response Framework," a new disaster response strategy that takes the place of the National Response Plan. Unlike its predecessor, the National Response Framework establishes a comprehensive, all-hazards approach that refocuses FEMA primarily toward more commonly occurring natural disasters. The National Response Framework includes terrorism, but only as one of many possible national emergencies. Federal officials have also emphasized the new response plan's flexibility and its application to incidents ranging from smaller, mostly local emergencies to large-scale terrorist attacks or catastrophic natural disasters. Another significant change in the new response plan is that it once again gives FEMA the leadership role in advising the president about disaster response in the event of a major natural or human-caused disaster or other emergency. Although the secretary of homeland security is identified as the principal official in charge of managing the federal response to a domestic disaster, FEMA's director serves as the primary advisor to the president when a disaster strikes and is the person who makes the urgent decisions about where to send or how to move resources.

Besides clarifying FEMA's mission and broadening FEMA's leadership role, the new response plan sets forth key response principles and structures and explains how municipalities, states, and the federal government will work together to provide a coordinated and effective national response to disasters. For example, the plan clearly provides for a bottom-up disaster response system, placing responsibility for responding to both natural and human-caused disasters on states and local communities. Local elected officials and emergency managers are to make the initial decisions about how to respond to an emergency and will be responsible for coordinating with private groups and nongovernmental aid organizations that may be able to help. The American Red Cross is designated as the lead organization for integrating the efforts of other national nongovernmental organizations that provide aid after a disaster. If the crisis cannot be contained by local resources, the locality can turn to the state for help. If conditions warrant, the state can activate the National Guard to help with rescue and relief efforts or to maintain order.

In the event of an emergency or disaster that local communities and states cannot handle on their own, the plan explains when and how the federal government can become involved. The main legislation authorizing federal disaster response is the Stafford Emergency Relief and Disaster Assistance Act (commonly called the Stafford Act), a law that allows the president to declare major national disasters or emergencies. Such a declaration allows the release of federal financial and other assistance to the state. Ordinarily, the request for a Stafford Act declaration will come from the state's governor, through FEMA, but in extraordinary circumstances, the president can act unilaterally to declare a major disaster or emergency without a formal state request.

Also, in certain situations, federal assistance can be offered without a presidential Stafford Act declaration. For example, the National Response Framework provides that prior to and during catastrophic disasters—such as large-magnitude earthquakes, other natural disasters that affect large population areas, or events such as a nuclear disaster—the federal government can act proactively to mobilize and deploy federal resources in anticipation of a formal request from the state.

FEMA began training for the new response plan in 2008, and the agency plans to conduct national simulated emergency exercises to determine the plan's effectiveness and the level of national preparedness. One issue that remains a concern, according to some experts, is how to improve communications within and between all levels of government during a disaster. Some states that are experienced in handling natural disasters have established effective computer and data management systems, but other states and localities still lack these technologies. How to improve communications and other elements of the national disaster response system are the focus of the viewpoints in this chapter.

FEMA Should Give States and Localities More Responsibility for Disaster Response

Patrick Roberts

Patrick Roberts is a postdoctoral fellow in the Program on Constitutional Government at Harvard University and an assistant professor in the Center for Public Administration and Policy in the School of Public and International Affairs at Virginia Tech.

Following a devastating hurricane, the Federal Emergency Management Agency [FEMA] is in crisis. Should it be abolished? Should emergency management responsibilities be given to the military? Returned to the states? Consider the descriptions from a post-disaster report:

Prior to the hurricane, "relations between the independent cities . . . and the county government were poor, as were those between the county and the state. . . . After the disaster these relations did not improve, which impeded response and recovery efforts."

When the hurricane first made landfall, the country initially reacted with a "sense of relief" because the "most populated areas had been spared the full brunt of the storm." After a few days, however, it became apparent that flood waters would swamp both urban and rural areas, leaving thousands without power, food, water, or the possibility of evacuation.

Compounding disasters—wind, floods, communications and power failures—led to catastrophe. While a large state might have had the resources to respond quickly, small states were overwhelmed. "They [small states] usually cannot hold up their end of the [federal-state] partnership needed for effective response and recovery."

Patrick Roberts, "FEMA After Katrina: Redefining Responsiveness," *Policy Review*, June/ July 2006. Copyright © 2006 by the Board of Trustees of the Leland Stanford Junior University. Reproduced by permission.

The severity of the disaster called into question the entire enterprise of federal involvement in natural hazard protection. "Emergency management suffers from . . . a lack of clear measurable objectives, adequate resources, public concern or official commitments. . . . Currently, FEMA is like a patient in triage. The president and Congress must decide whether to treat it or to let it die."

To address contemporary threats, [FEMA] must hone its natural disaster expertise and delegate authority for disaster response to states and localities.

FEMA's Decline

The above criticisms pertain not, as one might expect, to FEMA's recent woes in New Orleans following Hurricane Katrina but to the agency's dilatory [slow] response to Hurricane Andrew, which devastated South Florida in 1992. After Andrew, Congress gave the agency an ultimatum: Make major improvements or be abolished. With the advice of the emergency management profession and an enterprising director, James Lee Witt, the agency underwent one of the most remarkable turnarounds in administrative history. In 1997, [President] Bill Clinton called it one of the "most popular agencies in government." FEMA was well regarded by experts, disaster victims and its own employees. By 2005, however, the agency had once again fallen into ignominy.

Before issuing more cries for radical change at FEMA, reformers should look to the lessons of the agency's reorganization in the 1990s, which focused on natural disasters rather than national security. Its turbulent history shows that while the agency can marshal resources for natural disasters and build relationships with states and localities, it lacks sufficient resources to take on too many tasks. Today, FEMA faces a protean terrorist threat and an increasing array of technological hazards. To address contemporary threats, the agency must hone its natural disaster expertise and delegate authority for disaster response to states and localities. True, delegation runs the risk of returning to the days of ad hoc [unplanned] disaster preparedness, when government poured money into recovery without reducing vulnerability to disasters. Nevertheless, decentralizing response functions is the best way to prepare for an increasingly complex array of disasters, as the risks and strategies for recovery for different kinds of disasters vary so dramatically from region to region. . . .

From 1993 until 2001, FEMA was far better prepared to handle a catastrophic natural disaster than it was in 2005.

The Katrina Disaster

Hurricane Katrina showed that by 2005 the link between political support and speedy disaster response had been severed. Like any president, [George W.] Bush would have been best served by a FEMA that could respond effectively to natural disasters. But his administration wanted to take the agency in a new direction after 2001, subjecting its spending to greater accountability and including it in a larger organization devoted to security and terrorism preparedness. What caused FEMA to deteriorate so soon after having made a remarkable turn-around?

With so much blame to go around, the Katrina catastrophe was overdetermined. Some sections of New Orleans-area levees had been poorly constructed because of poor planning and botched contract work. State and local agencies had failed to plan adequately for the transportation, housing, and security that would be needed during an extended crisis. Once the hurricane bore down on New Orleans, local officials waited too long to issue an evacuation order that failed to account for the poorest residents, and state and federal agencies were too slow to provide rescue and recovery resources. When help finally arrived, it was poorly coordinated. Even at the height of FEMA's power under James Lee Witt, Katrina would have been a costly disaster. And yet from 1993 until 2001, FEMA was far better prepared to handle a catastrophic natural disaster than it was in 2005.

The agency had lost many of the elements essential to its turnaround of a decade earlier. Politically appointed emergency managers, including Witt, were replaced by appointees with little disaster experience, including, most famously, director Michael Brown, whose previous position had been with the International Arabian Horse Association. By 2003, departures, early retirement, and job dissatisfaction had sapped the agency's career force. The all-hazards, all-phases idea, too, was weakened when preparedness granting programs were moved out of FEMA into a separate office in the Department of Homeland Security. Turf wars put distance between the preparedness, response, and recovery offices.

The agency dug its own hole in the decade leading up to Katrina because of ever-greater public expectations for disaster relief.

Though FEMA could have used an infusion of experienced professionals, simply repeating the Witt recipe for reorganization would not have addressed the challenges of the twenty-first century. Career FEMA employees, like civil servants across government, began to retire in droves. At the same time, oversight committees began to worry that programs for mitigation and recovery lacked proper procedures to ensure that money was being spent wisely. Terrorism posed the greatest challenge. Witt initially had refused to take on more responsibility for terrorism preparedness because he thought the threat was too unpredictable for the agency to be able to address effectively. After 9-11, the country had no choice but to consider terrorism.

Most attempts to assign blame for Katrina focus on the Bush administration or poor state and local government response. While there is plenty of blame to go around, such a focus is misdirected. Reformers must attempt to understand FEMA's shortcomings in an effort to retool for the future.

The Problems of FEMA

The agency dug its own hole in the decade leading up to Katrina because of ever-greater public expectations for disaster relief, ever-greater specialization of preparedness and mitigation programs, and confusion about how terrorism fit into the all-hazards model. People had not always looked to the federal government for help during disasters, but during the twentieth century the level of assistance expected from the federal government before and after a disaster ratcheted upwards. The media broadcast images of FEMA agents rushing to disaster sites and FEMA relief workers helping communities rebuild, all of which reinforced the public's belief that the federal government owed the disaster-stricken public. Stories of federal relief were far more widespread than stories of investors and local governments choosing to invest in hurricane- or flood-prone areas, gambling that the federal government would bear the cost of rebuilding.

Terrorism complicated FEMA's efforts to respond to natural disasters, not by seizing resources formerly directed to natural disasters, but by adding new considerations to preparedness efforts. The numbers and amounts of grants devoted to emergency preparedness and natural disasters increased slightly from 2001 to 2005. Authoritative federal response plans invoked the all-hazards language, but the language was not reflected in organizational structures at lower levels of government. Enough resources flowed to natural disaster preparedness, but not enough attention was devoted to reconciling the different threats posed by terrorism and natural disasters, especially at the state and local levels. In a terrorist attack the FBI and law enforcement agencies take the lead because the disaster is a crime scene. In a natural disaster, however, the sole focus is rescue and recovery, tasks best left to emergency managers. When Katrina struck, states and localities had been crafting plans and procedures for terrorist attacks but in many cases had failed to refine plans for natural disaster response.

The Katrina disaster exposed the disconnect between preparation and response.

Meanwhile, as states and localities were struggling with how to address new concerns about terrorism, FEMA's success in developing mitigation and preparedness programs contributed to the unraveling of the all-hazards, all-phases principle. Over the years, FEMA had gained responsibility for increasingly differentiated grant programs, from the massive U.S. Fire Administration program to specialized grants for urban preparedness. But planning for floods requires a very different kind of expertise from the kind required for planning for fires or chemical spills. Increasingly specialized programs were given to more and more organizations with expertise in the relevant areas. The legislation creating the DHS scattered even more disaster preparedness granting programs throughout the department. As a result, Homeland Security Secretary [at the time of Katrina] Michael Chertoff moved grants for states and localities into a "one-stop shop" outside FEMA in the department. Granting programs became so specialized and organizationally separate from FEMA's response and recovery duties that emergency managers no longer thought in terms of "all hazards, all phases."

A Lack of Preparation

The Katrina disaster exposed the disconnect between preparation and response. Though the possibility of a catastrophic
hurricane and flood was a staple of local lore and expert reports, New Orleans failed to plan for Katrina with the urgency
and specificity that the response required. Plans were made
but never thoroughly rehearsed. To take one example, the National Response Plan, adopted in December 2004 with great
fanfare, gives the DHS broad authority during a catastrophe
to deploy "key essential resources" such as medical and search
and rescue teams, shelters, and supplies even without a request from state authorities. In the event of a catastrophe on
the scale of Katrina, the plan notes, "A detailed and credible
common operating picture may not be achievable for 24 to 48
hours (or longer) after the incident. As a result, response activities must begin without the benefit of a detailed or complete situation and critical needs assessment." The secretary of
homeland security possesses the legal authority to bypass normal disaster procedures to begin rescue missions and to deliver aid. The Katrina response, however, did not follow the
spirit of the plan. Despite federal authority to deliver resources soon after or even before a disaster, federal officials
complained that they were slowed because state and local
leaders did not request resources soon enough. At each level
of government, leaders failed to hash out their differences be-
forehand. As a result, officials ran into communications roadblocks that should have been uncovered before the disaster
struck.

Officials might have overcome roadblocks had they completed the Hurricane Pam training in 2004, a fictional exercise
to prepare for a category 4 hurricane in New Orleans. The
five-day exercise included emergency officials from 50 parish
[county], state, federal, and volunteer organizations. But
$850,000 into the exercise, FEMA's funding for Pam was cut,
and the key decisions that would vex authorities in Katrina

had not yet been made. No one had walked through how to handle communications failures, and there was no plan to organize evacuation or transportation and medical care immediately after the hurricane. The 121-page plan that emerged from the aborted exercise left many issues "to be determined."

If done well, completing the exercise might have forced Louisiana and New Orleans officials to have backup plans when communication broke down and federal help failed to arrive. The Pam exercise—in its early stages, at least—made disaster management seem too easy because the federal directions seemed so clear and certain. Walter Maestri, the emergency manager for Jefferson Parish, recalled in an interview with *PBS Frontline* (November 24, 2005) that federal authorities gave him blithe assurances after the exercise: "'This is what we're going to do. This is what you're going to do. This is what this one's going to do.' And the problem here that developed in Katrina is that the locals accepted that."

Emergency management has become too complex for a ... federal agency to coordinate. Only states and localities are able to weigh many-faceted concerns about a range of disasters.

The Pam scenario corresponded closely to the events of Katrina. Pam predicted that the 100,000 people who would fail to evacuate would be the most vulnerable to the storm. In New Orleans the majority of the approximately 900 dead were elderly people, many of whom lived in nursing homes and were never given an opportunity to leave town. In one such home, the staff had not expected the hurricane to be so severe and had ignored evacuation plans. When it was too late to transport the patients to safety, the staff fled for their lives, leaving 34 elderly patients to die in the floodwaters. More thorough preparation might have prevented these deaths and established lines of communication between nursing homes and emergency management coordinators. Local officials will always be the first responders, but federal agencies can help ensure adequate preparation before a disaster strikes by setting standards, funding preparation for rare events, and providing expertise.

FEMA's National Role

After 1993, FEMA reversed course and showed that it could be both effective and popular. The agency jettisoned its civil defense legacy and crafted a focus on "all hazards, all phases" emergency management that in practice emphasized natural disasters. When forced to address new concerns, however, from terrorism to the need for more accountable disaster relief, FEMA fell short. Emergency management has become too complex for a FEMA federal agency to coordinate. Only states and localities are able to weigh many-faceted concerns about a range of disasters and develop appropriate strategies. New Orleans, for example, faces far different hazards from those facing a northeastern industrial city with a similar population and demographics. In the wake of Katrina, reformers should resist the all-too-easy temptation to centralize control in the national bureaucracy and instead grant more power and responsibility for disaster preparedness, response, and recovery to agencies further down the federal ladder.

Reforming emergency management should proceed along three lines: focusing FEMA's tasks on emergency management, reviving the "all hazards, all phases" process, and giving states and localities more responsibility.

Still, decentralization poses the risk of returning to the days when emergency management was ad hoc and the federal government provided too much, too late to communities after a disaster struck. Disasters are, by definition, rare events that overwhelm the capacity of normal public institutions and practices. States and localities have little incentive to prepare for 100-year floods, yet such floods occur with alarming frequency in the United States.

Disaster management poses a paradox: If states and localities are to take more responsibility for disaster preparedness and response, the federal government must also take more responsibility for disaster preparedness. FEMA is best equipped to assemble best practices and encourage their adoption through the granting process so that communities reduce their vulnerability before disaster occurs. Reforming emergency management should proceed along three lines: focusing FEMA's tasks on emergency management, reviving the "all hazards, all phases" process, and giving states and localities more responsibility for disaster preparedness and response. . . .

The Challenge of Decentralization

FEMA's first major organizational challenge was to transform itself from a civil defense agency into a natural disasters agency. The agency faces a similarly formidable task today as it attempts to improve its natural disasters capability while not shortchanging the terrorist threat. While the agency can learn from the past, the solutions that worked in the 1990s do not translate directly to the present dilemma. FEMA faces the challenge of handing more responsibility to states and localities while taking a greater role in preparedness and mitigation than it has ever assumed. . . .

With the kind of leadership displayed during the 1990s, the agency can refocus emergency management agencies around the process of "all hazards, all phases" disaster management. Emergency management's greatest challenge is improving coordination among its parts as the number of disaster-related organizations grows along with the kinds of hazards, including terrorism and pandemic disease in addition to natural and industrial disasters.

Understanding emergency management as a process will be key to reform if two groups can be given more responsibility. First, states and localities must invest more in planning and in reducing vulnerability. The federal government can provide incentives for planning through the granting process, and it can provide disincentives for poor planning by making states and localities bear more of the cost of disasters. Second, if individual emergency managers understand their jobs as part of an "all hazards, all phases" process, they might make decisions based less on turf claims and more on a desire to reduce the damage that disasters wreak on citizens who deserve better from their governments.

FEMA Must Coordinate Better with Voluntary Agencies in Providing Mass Care After Disasters

U.S. Government Accountability Office

The U.S. Government Accountability Office (GAO) monitors the fiscal performance of the federal government, acting as a "congressional watchdog" on how the federal government spends taxpayer dollars.

Using lessons from the 2005 Gulf Coast hurricanes, the federal government released the National Response Framework (NRF) in January 2008. This report examines (1) why the primary role for mass care in the NRF shifted from the Red Cross to the Federal Emergency Management Agency (FEMA), and potential issues with implementation, (2) whether National Voluntary Organizations Active in Disasters (NVOAD)—an umbrella organization of 49 voluntary agencies—is equipped to fulfill its NRF role, (3) the extent to which FEMA has addressed issues with mass care for the disabled since the hurricanes, (4) the extent to which major voluntary agencies have prepared to better serve the disabled since the hurricanes, and (5) the extent to which FEMA has addressed issues voluntary agencies faced in receiving Public Assistance reimbursement. To analyze these issues, GAO reviewed the NRF and other documents, and interviewed officials from FEMA, voluntary agencies, and state and local governments.

U.S. Government Accountability Office, "National Disaster Response: FEMA Should Take Action to Improve Capacity and Coordination between Government and Voluntary Sectors," GAO-08-369, February 27, 2008. Reproduced by permission.

Working with Voluntary Organizations

FEMA and the Red Cross agreed that FEMA should be the primary agency for mass care in the NRF because the primary agency should be able to direct federal agencies' resources to meet mass care needs, which the Red Cross cannot do. The shifting roles present several implementation issues.

For example, while FEMA has enhanced responsibilities for coordinating the activities of voluntary organizations, it does not currently have a sufficient number of specialized staff to meet this responsibility. NVOAD has characteristics that help it carry out its broad role of facilitating voluntary organization and government coordination, but limited staff resources constrain its ability to effectively fulfill its role in disaster response situations. NVOAD held daily conference calls with its members after Hurricane Katrina, but these calls were not an effective means of sharing information, reflecting the fact that NVOAD had only one employee at the time of Katrina.

FEMA has begun taking steps in several areas to improve mass care for the disabled based on lessons learned from the Gulf Coast hurricanes.

Care for the Disabled

FEMA has begun taking steps in several areas to improve mass care for the disabled based on lessons learned from the Gulf Coast hurricanes. For example, FEMA hired a Disability Coordinator to integrate disability issues into federal emergency planning and preparedness efforts. However, FEMA has generally not coordinated with a key federal disability agency, the National Council on Disability, in the implementation of various initiatives, as required by the Post-Katrina Emergency Management Reform Act of 2006. The Red Cross has taken steps to improve mass care services for the disabled, but still faces challenges. For example, the Red Cross developed a shelter intake form to assist staff in determining whether a particular shelter can meet an individual's needs. However, Red Cross officials said that some local chapters are still not fully prepared to serve individuals with disabilities. Other voluntary organizations had not identified a need to improve services for individuals with disabilities, and we did not identify concerns with their services.

Communication Issues

FEMA has partially addressed the issues faced by local voluntary organizations, such as churches, in seeking Public Assistance reimbursement for mass care–related expenses after the hurricanes. At the time of the hurricanes, a key FEMA reimbursement program was not designed for a disaster of Katrina's magnitude, but FEMA has changed its regulations to address this issue. Local voluntary organizations also had difficulty getting accurate information about reimbursement opportunities. Key FEMA staff had not received training on reimbursement policies and sometimes did not provide accurate information, and some of the information on FEMA's Web site was not presented in a user-friendly format. FEMA has not addressed these communication issues.

Our recommendations from this work are listed below. . . . Status will change from "In process" to "Implemented" or "Not implemented" based on our follow up work. . . .

FEMA has not addressed [certain] communication issues.

Recommendation: To provide greater assurance that FEMA has adequate staff capabilities to support the agency's enhanced role under the NRF in helping coordinate with voluntary organizations, the Secretary of Homeland Security should direct the Administrator of FEMA to take action to enhance the capabilities of its Voluntary Agency Liaison (VAL) work-force, such as converting some Katrina VALs into full-time VALs able to work on the entire range of coordination issues with voluntary organizations.

Agency Affected: Department of Homeland Security [DHS]

Status: Implemented

Comments: DHS recognized the essential coordination role of VALs and is converting 12 VAL positions into permanent, full-time positions.

Recommendation: To provide greater assurance that FEMA has adequate staff capabilities to support the agency's en-hanced role under the NRF in helping coordinate with volun-tary organizations, the Secretary of Homeland Security should direct the Administrator of FEMA to take action to enhance the capabilities of its Voluntary Agency Liaison (VAL) work-force, such as increasing the number of full-time VALs.

Agency Affected: Department of Homeland Security

Status: In process

Comments: FEMA indicated that it has converted some on-call voluntary agency liaisons to permanent positions, but has not increased the number of full-time VALs.

Recommendation: To provide greater assurance that FEMA has adequate staff capabilities to support the agency's en-hanced role under the NRF in helping coordinate with volun-tary organizations, the Secretary of Homeland Security should direct the Administrator of FEMA to take action to enhance the capabilities of its Voluntary Agency Liaison (VAL) work-force, such as providing role-specific training to VALs, includ-ing providing them with information about Public Assistance opportunities and policies for voluntary organizations.

Agency Affected: Department of Homeland Security

Status: Implemented

Comments: FEMA has completed the development of a Voluntary Agency Liaison Handbook, complete with current guidance on how to perform their role effectively in disaster response and recovery operations. The Handbook has been piloted as a teaching module with two groups of FEMA VALs, and training will continue until all VALs have been trained. In addition, FEMA has completed the revision of a Volunteer and Donations Management course and plans to begin training VALs in the summer of 2008.

Recommendation: In light of FEMA's enhanced role under the NRF in helping coordinate the activities of voluntary organizations in disasters, the Secretary of Homeland Security should direct the Administrator of FEMA to provide technical assistance to NVOAD, as needed, as NVOAD works to improve its communication strategies.

Agency Affected: Department of Homeland Security

Status: In process

Comments: FEMA indicated that it is committed to providing technical assistance to strengthen NVOAD's information-sharing capability and that it will participate on NVOAD's Communications Standing Committee, but has not demonstrated that it has yet provided technical assistance for FY[fiscal year]08.

Recommendation: To ensure that the needs of individuals with disabilities are fully integrated into FEMA's efforts to implement provisions of the [Post-Katrina Emergency Management Reform] Act that require FEMA to coordinate with National Council on Disability (NCD), the Secretary of Homeland Security should direct the Administrator of FEMA to develop a detailed set of measurable action steps, in consultation with the NCD, for how FEMA will coordinate with NCD.

Agency Affected: Department of Homeland Security

Status: In process

Comments: FEMA indicated that it has taken several steps, including assigning a FEMA staff member responsibility for coordinating with NCD, but did not indicate that it had developed a detailed set of measurable action steps for FEMA/ NCD coordination for FY08.

Creating a More User-Friendly Web Site

Recommendation: To help ensure that voluntary organizations can readily obtain clear and accurate information about the reimbursement opportunities offered by the Public Assistance program, the Secretary of Homeland Security should direct the Administrator of FEMA to take action to make the information on FEMA's Web site about reimbursement opportunities for voluntary organizations more user-friendly. This could include developing a user-friendly guide or fact sheet that provides an overview of opportunities for reimbursement for facilities damage.

Agency Affected: Department of Homeland Security

Status: In process

Comments: FEMA indicated that it will continue efforts to improve the user-friendliness of Web-based information pertaining to reimbursement opportunities, but did not provide any specific actions taken in FY08.

Recommendation: To help ensure that voluntary organizations can readily obtain clear and accurate information about the reimbursement opportunities offered by the Public Assistance program, the Secretary of Homeland Security should direct the Administrator of FEMA to take action to make the information on FEMA's Web site about reimbursement opportunities for voluntary organizations more user-friendly. This could include providing contact information for organizations to get more information about Public Assistance program opportunities.

Agency Affected: Department of Homeland Security

Status: In process

Comments: FEMA indicated that it will continue efforts to improve the user-friendliness of Web-based information pertaining to reimbursement opportunities, but did not provide any specific actions taken in FY08.

Recommendation: To improve NVOAD's effectiveness in meeting its NRF information-sharing responsibilities after disasters, NVOAD should assess members' information needs, and improve its communication strategies after disasters. As part of this effort, NVOAD should examine how best to fund improved communication strategies, which may include developing a proposal for FEMA funding. To facilitate the implementation of improved communication strategies, NVOAD may want to consider strategies for increasing staff support for NVOAD after disasters, such as having staff from NVOAD member organizations temporarily detailed to NVOAD.

Agency Affected: National Voluntary Organizations Active in Disasters

Status: In process

Comments: We have requested but not received any information on NVOAD actions to address this recommendation for FY08.

A Good Communications Network Is Still Needed for First Responders

Sascha Meinrath

Sascha Meinrath coordinated the Community Wireless Emergency Response Initiative, a community group that volunteered time to help restore communications in areas hit by Hurricane Katrina. He also is the founder of CUWiN.net, a foundation that develops community-owned communication networks, and EthosWireless.com, a wireless consultancy focused on social justice.

Contrary to popular perception, the problem of disaster recovery is often not the lack of resources, but lack of coordination.

One key component to successful emergency response is a dynamic, direct and robust communications network—a structure the United States had been missing. Key decisionmakers turned a deaf ear to the problem until Hurricane Katrina made such an ostrich-stance untenable, and the United States had to learn the lesson the hard way. Yet [at the end of 2006], improvements have been incredibly modest. During the next major disaster, experts say we should expect more of the same—a lack of coherent, rapidly deployable, interoperable communications networks for first responders and the communities they serve.

In many ways, the state of U.S. disaster response is not too different from what we see in far less developed areas of the globe. Following the magnitude 7.6 earthquake that struck Pakistan, India and Afghanistan on Oct. 8, 2005, many problems faced by first responders were eerily similar to those experienced in Katrina's wake. According to one Indian IT [information technology] expert familiar with the situation, "The machinery of government had difficulty getting and sending even a handful of satellite phones for use in the devastated areas. I don't know if any of them have fully ready-to-move transportable (airliftable) satellite video uplinks, which would certainly be very useful. Similarly equipment for receiving remote-sensing imagery in real time and GPS [global positioning system]/location equipment [was lacking]."

One key component to successful emergency response is a dynamic, direct and robust communications network.

Jeff Allen, a consulting engineer currently working in Liberia with Médecins Sans Frontières [Doctors Without Borders], was a key member of [the nonprofit groups] Radio Response and the Community Wireless Emergency Response Initiative following Hurricane Katrina. Both groups developed and deployed critical telecommunications and network infrastructure in the hurricane's aftermath.

In terms of U.S. scenarios for emergency communications and disaster response, Katrina provided a sobering example of what works, what doesn't work, and the lessons we could learn from the ensuing massive communications meltdown. Allen's on-the-ground experiences helping to coordinate telecommunications disaster recovery were presented to the FCC [Federal Communications Commission] on March 6, 2006. . . .

What Worked, What Didn't

Generally speaking, hands-on investment in disaster preparedness is both sorely needed and relatively lacking. Designing networks to be deployed in advance is one of the most valuable lessons disaster recovery workers learned. Caching equipment and training recovery teams are also critical to these efforts. Yet more than a year after the largest natural disaster the U.S. has ever faced, little has been done to improve communities' preparedness.

Instead of improving communications, new emergency response systems often expand the gulf between responders and the communities they are supposed to help.

During disaster recovery, one of the most important elements is the organization of human beings. Thus, current initiatives to create separate infrastructures for "official" responders and the rest of the community are met with skepticism by those who have worked on the ground. "I have heard some vendors talking about municipal networks with VLANs [virtual local area networks] for public access and VLANs for public safety people," Allen said. They tend to treat the public access as an add-on, or as a luxury that can be turned off when bandwidth gets tight. That's lunacy. Giving people the tools to work together and solve their own problems is way more powerful than giving 20 police cars full motion video over a wireless Ethernet system. Humans need low bandwidth and existing collaboration systems hosted out in the network to organize to help themselves.

Instead of improving communications, new emergency response systems often expand the gulf between responders and the communities they are supposed to help by creating additional technological barriers to shared use.

And while the technology stories that emerged from Katrina often focused on the glamour of certain services . . . , tried and true applications like instant messaging (IM), private chat rooms, and Web access actually worked best. "VoIP [voice over Internet protocol, an Internet telephone service] is OK for networks that are fully controlled, and whose topology and capacity are well planned," Allen said, "but it's unusable on the kind of agile, fluid and low-bandwidth network you find in a disaster area."

Unfortunately more municipal networks are being built using both hierarchical and centralized infrastructures—all but guaranteeing they will be more prone to failure during a disaster. "A closed system that depends on a proprietary [requiring a license] configuration server would be dead in the water when the configuration server lost power (a common occurrence in a disaster area)," according to a Radio Response report.

During emergency response, one should expect a lack of advanced technological know-how from most emergency responders.

Likewise, newer communications technologies, some of which were first deployed during the post-Katrina disaster recovery, often turned out to be nonfunctional. Though many press releases from major corporations heralded the successes of WiMAX [Worldwide Interoperability for Microwave Access, a wireless technology] and mesh [impromptu networks that work through many small nodes], first-experiences were different. "In Mississippi I did not see any indication that mesh technology works," Allen said. "I don't know why not. The arguments for it are compelling and the technology has had enough time to be stable, it seems. It might just be that deploying enough working nodes and keeping them working is the problem."

Disaster recovery nationwide is often facilitated through incident command systems (ICSs) that respond to everything from a single house fire to a Katrina-sized event. While ICS training is accessible online and through the Federal Emergency Management Agency (FEMA), few communities have made this training available to emergency responders and interested residents. One important element of ICS is the emergency support facilities (ESF), which consist of numerous divisions that each deal with a different facet of disaster recovery. Though the goal of these divisions is to coordinate responses, often in-fighting and personal politics led to perception of them as a barrier to effective on-the-ground recovery. For the Community Wireless Emergency Response Initiative, this meant interfacing with ESF-2 (Communications), ESF-15 (Volunteer Coordination) and ESF-5 (Facilities), and getting approval from each for different elements of their work.

During emergency response, one should expect a lack of advanced technological know-how from most emergency responders. As documented for the FCC in the Radio Response report, "it seems that most people who handle radios for emergency operations do not understand electronics, physics or RF [radio frequency] propagation . . . they consider any nongovernment use of RF equipment a threat to their turf. People coming from this point of view are rarely swayed by facts or by regulations." Time and again, officials dismiss innovative solutions to emergency communications because they don't understand the technologies.

Allen laid out four simple recommendations to the FCC to help facilitate successful telecommunications setup during disaster recovery, proposals that a municipality of almost any size could easily establish:

- Preplan the network architecture, including "what if" scenarios for how to modify the architecture in response to a situation.

- Maintain a cache of preconfigured and known-good hardware.

- Take part in real-life drills using the real cache hardware.

- Have at least one experienced staff member on call to provide leadership and continuity during an actual disaster response.

The Radio Response team recommends easily deployable hardware. "The devices have to act like simple appliances. Configuration should be via Web user interfaces. If we are to have a dynamic routing system, it must work in the home networking context." In other words, plug-and-play, off-the-shelf hardware is often preferable to far more specialized (and often more expensive) solutions. Unfortunately numerous Community Wireless Emergency Response team members found themselves spending too much time troubleshooting donated equipment, and worrying about relatively unimportant logistical details and potential legal issues.

Network engineers sporadically found that the available hardware equipment was unstable, broken or in need of software upgrades. In addition, maintaining an up-to-date, "known-good" equipment cache is critical because much of the equipment on the market is problematic out-of-the-box. "The quality of the engineering of the software—and to a lesser extent, hardware—is very low in these types of devices," Allen said. "Software bugs are very common, and unless you are using a particular 'blessed' version of the firmware, behavior is far from predictable."

Furthermore, during the Katrina disaster recovery, obtaining legal Windows licenses for refurbished computers installed in refugee centers proved impossible, and responders often resorted to using illegal copies of the software. As a result, one of the recommendations made to the FCC was to utilize nonproprietary systems whenever possible. In the case of computer operating systems, responders recommended Linux LiveCDs, which let a computer boot directly from the CD itself, and allow users to quickly burn more operating system disks as needed. "Fetching, burning and running a LiveCD is practical in a disaster context," wrote Allen. "Debugging is not. We wasted a significant amount of time with configuration errors. It is easy to make them in the context we were working in, and it was exceptionally difficult to find them and fix them."

In disaster recovery, the "official playbook" doesn't always conform to on-the-ground realities. "Both FEMA and Red Cross depended to a huge extent on telephone service working," Allen said in his report. "Their behavior in this regard was strange, as it seemed to disregard the reality that close to 100 percent of the victims from Hancock County were without reliable personal telephone service." Once emergency responders restored Internet connectivity, often well ahead of phone service, storm survivors found even more disturbing practices from the major players. For example, FEMA's Web site required a relatively recent version of Internet Explorer (IE) to complete forms to receive federal aid. Users who ran Linux, Macintosh or other IE-incompatible operating systems couldn't apply for FEMA assistance online.

Better Today than Tomorrow

After Katrina, Community Wireless Emergency Response Initiative team members like Mac Dearman, Will Hawkins, Joel Johnson and Paul Smith focused on coordinating and setting up telecommunications infrastructure as quickly as possible in incredibly chaotic environments.

Internet infrastructure is surprisingly low on the priority list during disaster recovery.

Enthusiasm, adaptability and a MacGyver-esque ethos were often the most important elements to the successful completion of the day's tasks. This often means telecommunications recovery efforts need to revisit previous work sites to make network upgrades. Allen put it this way, "You have to learn that when you are operating day by day on what could charitably be termed a 'good plan,' you must schedule time later for rework, to incorporate the unknowns the 'good plan' glossed over. This is true in all network design, I think, but it is a bigger deal when the cycle time is so short; a network built last week might be ready for significant rework this week. This is a common problem in the emergency management context." During disaster recovery, one often doesn't have the luxury to wait for orders or directions from higher-ups. In addition, the authorities don't always know how to get things done better than the local community does. "If you expect to get direction—or even accurate intelligence—from the authorities, you'll be disappointed," said Allen.

Internet infrastructure is surprisingly low on the priority list during disaster recovery. As the Radio Response report stated, "The priorities are transport (without which you can't move resources to solve any of the other problems), then communications, then survival commodities like water and (later) food. Communications is a very high priority, but the needs are met with a small set of linked VHF [very high frequency] repeaters and stand-alone satellite connections, not with an Internet distribution network." Though reprioritization of this critical resource will eventually happen as government agencies realize the importance of Internet connectivity, for now, most communities end up cut off from all Internet-reliant services for long periods of time whenever disaster strikes.

Public access isn't the only reason to restore Internet connectivity as fast as possible. If post-Katrina network activity statistics are any indicator, the networks set up by the Community Wireless Emergency Response Initiative saw an equal amount of usage from disaster relief workers themselves. Responders often used the network to communicate with their home bases, since cell phone coverage was often spotty and landline communication completely nonexistent. Workers also used the network to depict the disaster's impact on daily life. "Individuals used the Internet connection to explain what they were experiencing to friends back home," Allen said. "They sent out e-mail to worried parents and posted to blogs. Sharing their experiences like this helped attract more volunteers and resources to get the job done."

The most common use of computers connected to these networks was Web-based e-mail (Hotmail, Yahoo, Gmail, etc.). However, users also downloaded IM clients to communicate in real time with friends and family. Additional examples of network usage included:

- coordinating with supporters back home to bring more supplies and volunteers;

- remotely accessing workplace computers;

- paying bills;

- searching for new jobs;

- complaining about current conditions;

- posting news and commentary; and

- checking prices on eBay for selling salvaged collectibles to raise cash for house repairs.

Solutions and Lessons Learned

Many in the emergency management community recommend public investment in open standards and nonproprietary, interoperable technologies. While VoIP and other services are exciting new tools, the most critical applications are the latently strong ones like e-mail, Web access and IM. "Volunteer community teams" are . . . fundamental to disaster recovery. Data collection and management are also important, and critical information should be easily accessible to responders:

- All geographical information—maps, GPS readings, driving directions—is needed by responders. Often, fundamental assumptions need to be shifted. For example, relying on street maps is useless when signs are blown down, and referring to landmarks is sometimes the best option.

- Contact information for team members, current and prospective key customers/survivors, government and aid organization liaisons, and contractors is often overlooked or difficult to access. Best contact mediums also shift often during disaster response.

- Network information—IP [Internet protocol] allocation plans, currently assigned network numbers, network diagrams, passwords, administrator contacts—is also rarely accessible or systematically tracked. Ensuring that important communications network information is current and available will help save time—and avoid headaches. Redundant and robust networks such as satellite links, wireless, landline connections require more information than many traditional single-medium system architectures.

Many in the emergency management community recommend public investment in open standards and nonproprietary, interoperable technologies.

- Equipment information—notes on configurations, hardware and cabling, vendor manuals and inventory—is vital. One of my favorite stories is about "Brent" who brought boxes of one-gallon heavy-duty Ziploc bags to save storage space and reduce the time spent searching for the correct wall plug or cable by letting responders package equipment—e.g., router, Ethernet cable, power brick and manuals—into transparent, waterproof, easily opened units.

- Documentation—timelines, photos, press, video, notes—is often downplayed during disaster recovery.

Yet learning from previous initiatives is vital to continued improvement of recovery efforts.

Most disaster recovery responders said priority is often placed on "outside experts" and "professional services," but the successful mobilization of local community assets is also critical to disaster recovery. For example, many community teams are perfectly situated to help build and interconnect telecommunications networks, and often have the social capital necessary to bring diverse constituencies to the table.

Practice, preparation and preplanning were hailed as fundamental to facilitating smoother operations during the chaos of disaster response. "Help communities take care of themselves," Allen said. "It's respectful of people's need to be involved in their own recovery. It gives you huge leverage with small investments, and it lets the best aspects of American culture—like teamwork, ingenuity and giving—overcome the worst situations."

FEMA Is Still in Need of Reform

Mitchell L. Moss

Mitchell L. Moss is a professor of Urban Policy and Planning at New York University's Wagner Graduate School of Public Service.

The California wildfires [in 2007] have reinforced the need to recognize that manmade and natural catastrophes of unimaginable scale are now normal parts of everyday life. Sound public policies to cope with disasters require skilled leadership and a system designed for 21st century threats. While the Federal Emergency Management Agency [FEMA] has been strengthened since Hurricane Katrina when it failed to meet the needs of the Gulf Coast, and FEMA's improvements were evident in the response to the California wildfires, the underlying federal disaster response system is still in need of reform. Its mission is seriously hampered during large-scale events by regulations designed to respond to smaller scale disasters.

Outdated Laws and Regulations

The Stafford Act—the law at the heart of the federal government's disaster response and recovery activities—is simply not designed to cope with catastrophic disasters. Passed in 1988, the Robert T. Stafford Disaster Relief and Emergency Assistance Act authorizes federal disaster assistance to residents, businesses and local governments. However, the Stafford Act was written to deal with the smaller scale disasters that are most common within the United States such as tornadoes, blizzards and floods. No distinction is made within the act be-

Mitchell L. Moss, "Out of Scale: We Need a Disaster Policy Sized to Our Catastrophes," *San Francisco Chronicle*, December 26, 2007. Reproduced by permission of the author.

tween an October 2006 blizzard that hit Buffalo, N.Y., the devastation wrought by Hurricane Katrina or the recent California wildfires. Under the act, all were declared "Major Disasters." The act should be updated to recognize a new level of response beyond "Major Disasters," one designed specifically for large-scale and devastating "Catastrophes."

While the Federal Emergency Management Agency has been strengthened since Hurricane Katrina ... [the] federal disaster response system is still in need of reform.

Regulations designed to address small-scale disasters have become bureaucratic roadblocks following a catastrophe. After Hurricane Katrina, the city of New Orleans was forced to lay off nearly half its workforce because there was not enough cash on hand to meet payroll obligations. The act authorizes federal disaster assistance to cover the overtime of local government workers for work related to responding to the disaster, but not regular salaries. The Stafford Act should be amended to ensure that communities suffering from a catastrophe do not face the duel challenge of overwhelming devastation and a bankrupt local government.

Other Changes Needed

Federal disaster policy assumes that financial assistance to business and individuals should come from private insurance before government steps in. Unfortunately, the devastation that follows a catastrophe can overwhelm insurance companies and result in extensive delays for insurance payments. Government policy must continue to encourage insurance coverage so that insurance companies, not taxpayers, absorb the bulk of the costs following a catastrophe. However, the Stafford Act and its regulations should be modified to allow cash assistance—which is capped at $28,800 per household, and further subdivided with caps on repairs, temporary housing assistance and other items—to flow to qualified homeowners and renters following a catastrophe without having to provide proof of insurance coverage. This will provide households with immediate assistance to start the rebuilding and recovery process. This assistance could later be reimbursed to the government when insurance coverage is received.

The Stafford Act—the law at the heart of the federal government's disaster response and recovery activities—is simply not designed to cope with catastrophic disasters.

Following a catastrophe, the resumption of electric, water, sewer and telecommunications services is critical. Lives can be saved and suffering minimized with functioning utilities. However, utility workers are not treated as emergency responders under the Stafford Act. While utility crews in Southern California were able to quickly restore power to San Diego after fire damaged the Southwest Power Link, in New Orleans, BellSouth had to shut down telephone lines that were still working after Hurricane Katrina because the BellSouth workers were not recognized as "emergency responders" and therefore were not eligible to receive security escorts and "priority" access to food, fuel, water and shelter from the federal government. The Stafford Act should be amended so that utility workers are recognized as "emergency responders" after a catastrophe.

In 2006, Congress passed a law mandating that the head of FEMA demonstrate emergency management ability and would have at least five years of executive leadership experience. This was an important first step that has already yielded improvements—competent management enabled FEMA to respond to California's wildfires far more effectively than it did to Hurricane Katrina. But more change is needed. Comprehensive Stafford Act reform must be passed to ensure that the federal government has the capacity to respond to modern threats and the catastrophes that we have become accustomed to in the 21st century.

FEMA Should Not Be Given More Power

Michael Hampton

Michael Hampton is the editor and publisher of Homeland Stupidity, a Web site critical of government excess and mismanagement.

When the next hurricane threatens to strike, how will you get the news? For that matter, will you survive? Some want to give the Federal Emergency Management Agency [FEMA] even more authority over disaster response than it already has, even while it struggles to modernize the country's emergency alert system.

FEMA has gotten a virtual free pass for the last two years; since Hurricane Katrina struck New Orleans down in August 2005, there have been no hurricanes or other disasters of any comparable size.

Yet some claim that FEMA's failures in responding to Katrina derive from it not having enough power under the law to accomplish its mission. Senate lawmakers are currently [in May 2008] drafting legislation to update the Stafford Act of 1988, under which FEMA has responsibility for disaster response, which Senate staffers say does not cover catastrophic events like Hurricane Katrina.

Mitchell Moss, the Henry Hart Rice Professor of Urban Policy and Planning at New York University and an investigator at the center, said of the Stafford Act, "Despite good intentions, it doesn't work. Congress is always having to work around its limits."

Michael Hampton, "Does FEMA Need More Power?" Homeland Stupidity, May 18, 2008. Reproduced by permission.

Among the limitations Moss cited, the law caps federal loans to state and local governments to offset lost tax revenue following a disaster at $5 million—a wholly inadequate figure. In 2002 and 2003, for example, New York City lost nearly $3 billion in tax revenues following the Sept. 11, 2001, terrorist attacks. After Katrina struck, New Orleans had to lay off almost half of its workforce—about 3,000 employees—because the city didn't have enough cash to pay them (the law allowed the federal government to reimburse the city for employee overtime, but not for the salaries themselves).

FEMA forced [New Orleans residents] to suffer and allowed them to die by . . . keeping out rescue workers and relief supplies . . . and tying victims up in red tape.

Not only did the city face overwhelming devastation, but with its tax base destroyed it had no way to pay employees when it needed them most, Moss said.

In addition, the law prohibits federal assistance to utilities except if those utilities are publicly owned or nonprofit. This was an impediment to New Orleans regaining phone service after Katrina because in the lawless interlude that followed, BellSouth could not provide security for employees needed to maintain service, and the federal government was prohibited from assisting, Moss said. Utility workers should be considered "emergency responders" in the aftermath of a disaster or catastrophic event, he added.

Nowhere in the discussions, unfortunately, is any mention made of the real reason why so many people suffered and died in New Orleans. FEMA forced them to suffer and allowed them to die by, among other things, keeping out rescue workers and relief supplies, not knowing what they're doing, and tying victims up in red tape. Oh, did I mention wasting taxpayer money?

The Need for a New System

It gets better. President [George W.] Bush in 2006 ordered the Department of Homeland Security [DHS] to modernize the nation's emergency alert system, and DHS gave the task over to FEMA. Two years later we've seen nothing but the occasional prototype and pilot project and a whole lot of talk, but the so-called Integrated Public Alert and Warning System [IP-AWS, which FEMA was given an executive order to develop as a replacement for the Emergency Alert System] is no closer to reality.

The House [of Representatives] Homeland Security subcommittee on emergency communications, preparedness and response held hearings on the state of the IPAWS system, with subcommittee chairman Rep. Henry Cuellar (D-Texas) calling for FEMA to explain why it hasn't fully implemented the executive order.

In a Feb. 19 [2008] filing with the FCC [Federal Communications Commission], less than two months before the commission adopted technical rules for the commercial mobile alert system, FEMA assistant administrator Martha Rainville said FEMA lacked statutory authority during non-emergency periods to be involved with critical components of the commercial mobile alert system, including aggregator and gateway functions as well as the trust model, when warnings are issued by non-federal agencies.

"We cannot do everything at once so later this year we are rolling out the first increment to support digital alerts," Rainville said in written testimony. "Later on, we will roll out additional increments to support risk-based alerts, non-English language alerts and alerts for special-needs communities."

The country's existing Emergency Alert System is an audio and text only broadcast distributed over television and radio networks. The IPAWS system would "support audio, video, text and data messages sent to residential telephones, to Web sites, to pagers, to e-mail accounts and to cell phones," Rainville said.

Of course, if you think those alerts are coming to your cell phone any time soon, think again. Rainville said that FEMA doesn't have statutory authority to implement parts of the system.

If you think [disaster] alerts are coming to your cell phone any time soon, think again.

In the FCC's commercial mobile alert ruling on April 9 [2008], Chairman Kevin Martin said it would have been better if a federal entity were in place to oversee alert aggregator and gateway functions. Commissioner Michael Copps was more critical of FEMA in his statement, triggering an angry response the following day.

"It is unfortunate that Commissioner Copps chose to question FEMA's role and responsibility without first talking with the agency's administrator before making his provocative comments," said FEMA in a statement. The statement said Copps mischaracterized FEMA as an unwilling partner in the process to reform the nation's public warning system. FEMA also accused Copps of failing to mention . . . FEMA's apparent lack of clear legal authority during non-emergency periods to manage the commercial mobile alert system.

The system uses the standards-based Common Alerting Protocol [a data format] internally, but no provision has yet been made to provide the data to the public.

Forget FEMA

FEMA is the agency, some people think, that somehow needs *more* power and authority in order to respond effectively to disasters. It seems that they've misused the power and authority they already had. Giving them more power and control simply will mean more misuse of power, more widespread impact of erroneous emergency messages, and more disaster victims needlessly suffering and dying.

The bitter irony of Hurricane Katrina is that fewer people would have died and New Orleans would have recovered more quickly if the federal government had not responded in any way.

Don't you feel safe now? You shouldn't. Forget about Homeland Security and get yourself and your family really ready for the next disaster.

Greater Government Investments Are Needed to Provide Effective Disaster Response

Donald Cohen

Donald Cohen is executive director of the Center on Policy Initiatives, a San Diego–based research and policy center.

Despite a tragic history of regular disasters—earthquakes, hurricanes, bridge collapses and fires—Americans seem unwilling to prepare for the inevitable. This may have something to do with the forty-year conservative assault on government and the resulting skepticism about things that can't be justified as fighting terrorism.

But local agency response to San Diego's wildfires [in 2007] shows that despite inadequate resources, government on the ground can, in fact, be good. The firefighters, police and other emergency personnel performed efficiently and heroically. Public officials at the local and state level worked well together to coordinate the firefighting, rescue and relief efforts.

San Diego's new, widely praised "reverse 911" system, which made thousands of evacuation calls to get residents out of harm's way early, was developed with Urban Area Security Initiative (UASI) Homeland Security funds. Even FEMA [Federal Emergency Management Agency]—the federal agency that bungled the response to Katrina—showed up quickly and began collecting relief applications.

Local government response to the San Diego fires is a stark rebuke to the claim that government is inherently incompetent—or even unnecessary. But while thousands of volunteers stepped up to help, we cannot plan a primarily volunteer response to major disasters, any more than we can expect volunteers to build our roads or collect our trash. This is why we have government. The only question is whether our public agencies will be adequately funded, equipped and staffed to do the things we need them to do.

A Bigger Story

What occurred in San Diego tells a bigger story. Preparing our nation for future disasters requires government at all levels to provide the resources we need to save lives and property. We need local, state and national leaders who can articulate a sense of common purpose beyond fighting terrorism—a vision of a nation that builds on hope, not fear. And they have a responsibility to identify the resources we need and to mobilize the public to pay for them.

Preparing our nation for future disasters requires government at all levels to provide the resources we need to save lives and property.

Unfortunately, as we head into a wide-open presidential election year [in 2008], none of the candidates are articulating an adequate response to our fundamental challenges. GOP [Republican] rivals parrot the same old, increasingly irrelevant formula of cutting taxes and dismantling government. And Democrats have yet to go far enough to articulate the critical role of public investment. So far, there are no bold proposals to build a twenty-first-century infrastructure that better prepares us for both public disasters and the daily crises facing American families.

A Lack of Political Will

The tragic losses of life and property from the San Diego wildfires, Hurricane Katrina and the [2007] Minnesota bridge collapse expose a troubling neglect of our nation's infrastructure. Together, they should be a wake-up call. We have the technology and know-how to protect ourselves. We just lack the political will.

Many disasters, such as hurricanes, fires and earthquakes, are predictable. We may not know when they'll arrive, but we know they will sooner or later. Other disasters—such as bridge collapses—seem more random, but we know that if bridges aren't regularly inspected and repaired, some will eventually collapse.

How many lives and how much property damage could have been saved if San Diego had been better prepared for the inevitable fire?

[In 2006], San Diego fire chief Jeff Bowman resigned, frustrated by the city's failure to pay for enough firefighters, stations and equipment to serve a growing population. He repeatedly used words like "ill-equipped" and "understaffed" to describe his department. Indeed, after San Diego's previous major fires—the Cedar and Paradise fires [of 2003]—Bowman and others documented that San Diego simply did not have enough equipment and personnel to meet these kinds of challenges.

San Diego leaders have long known about the city's underfunding for critical infrastructure. A 2005 study by the Center on Policy Initiatives, a [San Diego] nonprofit think tank, revealed that San Diego's per capita spending on fire protection is the third lowest among major California cities. In number of firefighters per 1,000 residents, San Diego ranked dead last. According to national firefighting standards, a city San Diego's size (1.25 million) should have at least twenty-two more stations and 400 more firefighters than it currently has. And in San Diego's desert climate, resources should be even larger than what standards suggest. The city's budget director recently estimated a long-term unmet need of $478 million to get to full firefighting capacity and an additional $40 million needed in the city's $1.1 billion annual operating budget.

As with the neglect and deterioration of the levees in New Orleans's hurricane-prone region, San Diego let fundamental foundations of emergency preparedness and healthy economic growth fall into disrepair. Public employees and agencies have shown they'll do their best when duty calls. In the end, though, things we value—like fire protection—aren't free.

The Need for Investment

The crises in San Diego, New Orleans and Minnesota are not unusual. They are simply vivid examples of how our chronic public underinvestment is harming the country and putting too many people at risk. Our lack of preparedness in all three disasters meant that people suffered more damage and hardship than they had to. Across America, in rural and metropolitan areas alike, infrastructure is crumbling, public agencies are understaffed and equipment is outdated.

There's also an opportunity to expose deeper levels of disinvestment in San Diego and across the country that cause day-to-day human-made crises in families, weaken our economy and leave us unprepared to deal with the potentially catastrophic impacts of global warming. We watch dramatic pictures of fires and floods on the news, but we don't see the invisible, though equally devastating, crises of the millions living without health insurance. We don't see the lost productivity of young people struggling in underresourced schools or those unable to afford a college education.

We need a national public works plan to help prepare our communities for both major disasters and the day-to-day disasters that undermine our country's strength and our future. We need to maintain and rebuild our highways, roads, bridges, tunnels, sewers and water delivery systems—the foundations that make commerce and productive regional economies possible. Now more than ever, we need to invest in renewable energy and conservation technologies to offset the environmental and public health impacts of global warming. And we need more schoolteachers, nurses and vocational skills educators to guarantee that the next generation of Americans are educated and trained for citizenship and the workforce. This is what government is supposed to do and, when provided adequate resources, does well.

Even tax-averse San Diego voters may soon recognize that there's no free lunch; we must take as good care of our common public needs as we do of our own homes.

These investments—along with buying more fire engines in Southern California, reinforcing levees in New Orleans and rebuilding deteriorating bridges in Minnesota and elsewhere— define the essentials of an American infrastructure that only government can address.

The Government Should Create a National Disaster Insurance Fund

Frank McEnulty

Frank McEnulty is an American businessman who ran in the 2008 presidential election, both as a vice presidential nominee of the Reform Party of the United States of America and as the presidential nominee of the New American Independent Party.

There was a terrible tornado in Kansas [in 2007]. I haven't seen total devastation on that scale in America since I was living in Florida and Hurricane Andrew went through the southern part of the state. No one, no community can fully prepare for a disaster of that magnitude, and it requires the help of local, state and Federal officials to help get things back to normal.

However, it has long been my belief that politicians secretly welcome natural disasters for the opportunity it gives them to pander to the electorate. While they may certainly care about the people affected by these tragic events, politicians also use them to be seen as caring and to throw our money around.

For years I've thought that there should be a more rational and organized response to natural and other disasters than the current system of knee-jerk politicians swooping in and doling out our money in an inefficient and wasteful way.

[If I were] President, I would propose a national disaster insurance fund be established to deal with these horrible situations. Although, as they say, the devil is in the details, a rough outline of how I believe the plan should work and be set up would be as follows:

Frank McEnulty, "The Federal Government and National Disasters," Mcenulty.word press.com, 2008. Reproduced by permission.

There would be a specific set of disasters covered under the insurance. For example, the policies would cover losses from earthquake, flood, hurricanes and tornadoes.

1. Current owners of property would have 5 years to decide to get into the program. After the 5 year transition period was over, if you didn't have the insurance, there would no longer be any Federal assistance to individual property owners in the event of a natural disaster.

2. For buyers of new property, they could decide to buy the coverage anytime after buying their property. However, if they did not purchase coverage, they would have no right to Federal assistance in the event of a natural disaster.

3. The policies would be sold through existing insurance companies, much as earthquake coverage is currently sold in California.

4. No one would be forced to buy, but I would assume, much like fire insurance, that anyone who has a mortgage on their home or other property would be required to do so by their lender if they lived in a zone where there was the potential for any of these major disasters.

5. Prices would be kept reasonable by two key facts. First, there would be a large number of property owners involved in the pool which would spread the risk. Between those property owners in hurricane, earthquake, flood and tornado zones quite a bit of the country and a tremendous number of properties would be potential buyers. Second, the Federal government could subsidize the coverage—the subsidy coming from what is not given out each year in emergency disaster funding.

Making People Act Responsibly

This program would accomplish several things.

171

First, it would make people responsible for doing what is necessary and proper to protect their property in the event of a catastrophic loss caused by a natural disaster.

Second, it would establish a true fund to pay for these events.

Third, it would allow people in certain parts of the country to know that they aren't underwriting the cost of the rebuilding of areas where these things always happen. Whether it is hurricanes in the Southeast, tornadoes in the Midwest or earthquakes in California, why should someone in Arizona or Idaho (where natural disasters are fairly few) be required to contribute their tax dollars to continually rebuild beach front property or homes in "tornado alley"?

Finally, it would take away the ability of the politicians to use our money to further their own political gains—at least in this area of life. . . .

[A national disaster insurance fund] would make people responsible for . . . protect[ing] their property in the event of a catastrophic loss caused by a natural disaster.

If you have property, it is your responsibility to see that you have the proper insurance coverage to ensure that if something tragic happens you have the coverage to rebuild your house. I believe an insurance program of this kind is the government's "responsible thing to do".

The Government-Funded Disaster Safety Net Should Be Replaced with Free-Market Policies

Thomas A. Bowden

Thomas A. Bowden is an analyst with the Ayn Rand Institute, a public policy organization in Irvine, California, that promotes individual rights and free-market capitalism.

In a speech from New Orleans [in April 2008], Republican presidential candidate John McCain lashed out at the Bush administration for its response to Hurricane Katrina. McCain's remarks, which appeared calculated to make disaster relief a key campaign issue, revived harsh memories of the savage storm that inundated the Mississippi Delta in late August 2005, leaving more than 1,800 people dead and causing widespread property damage.

Although the floodwaters long ago receded, government officials are still counting the disaster's costs. Earlier this year [2008], the U.S. Army Corps of Engineers disclosed that 489,000 claimants are seeking damages caused by poorly designed levees. Of those claimants, 247 want more than $1 billion each, including one whopper for $3 quadrillion (a stack of a quadrillion dollar coins would reach beyond Saturn).

The tax dollars spent resolving those claims will augment the tens of billions already paid to restore and repopulate New Orleans, a below-sea-level bowl situated precariously amidst a lake, a major river, and a gulf, in a known path for hurricanes.

Disasters can sometimes shock a nation into questioning entrenched practices. But Hurricane Katrina, perhaps the worst natural disaster ever to befall America, has failed to spark serious challenge to long-standing government policies that actively promote building and living in disaster-prone areas.

The Katrina tragedy should have called into question the so-called safety net composed of government policies that actually encourage people to embrace risks they would otherwise shun—to build in defiance of historically obvious dangers, secure in the knowledge that innocent others will be forced to share the costs when the worst happens.

Without blaming the victims for having followed their own government's lead, it is time to question whether those policies should continue.

The History of the Safety Net

The first strands of today's safety net were spun in the nineteenth century, as the Army Corps of Engineers shouldered the burden of constructing and maintaining levees and other flood controls along the Mississippi River. From then to now, Congress and the states have responded to each new flood by installing newer, higher, and stronger barriers at public expense, as if the preservation of a city like New Orleans in its historical location were a self-evident necessity.

The Katrina tragedy should have called into question the so-called safety net composed of government policies that actually encourage people to embrace risks.

Throughout the twentieth century, new strands were woven into the safety net, first in the form of loans to disaster victims, then by direct grants, infrastructure repairs, loan guarantees, job training, subsidized investments, health care, debris removal, and a host of similar rehabilitative measures.

In 1968, the National Flood Insurance Program began supplying subsidized coverage for structures and their contents in flood-prone areas. Similar state-subsidized insurance programs arose for hurricanes in Florida and earthquakes in California. In 1978, the Federal Emergency Management Agency was created to coordinate the increasingly complex job of government disaster response.

At each juncture, more aid was funneled to disaster victims without serious challenge to the wisdom of encouraging people to occupy vulnerable locations.

In response to Mississippi floods, Florida hurricanes, and California earthquakes, the number of major disaster declarations almost doubled from the 1980s to the 1990s, from an annual average of 24 up to 46. At century's end, Congress was paying an average of $3.7 billion a year in supplemental disaster aid, with state taxpayers contributing many millions more. As of August 2007, Katrina relief alone had cost federal taxpayers $114 billion.

By gradual steps, this disaster safety net became part of the legal landscape, taken for granted by private investors and owners deciding to undertake new projects or rebuild storm-damaged areas. Relief programs—by minimizing, disguising, and shifting the real risks of defying natural hazards—became an active force distorting private decision-making and inviting even worse future tragedies.

Thus if a pre-Katrina Mississippian asked himself, "Should I build my house 10 feet above sea level, a quarter-mile from the Gulf Coast?" the answer came back: "Sure, why not? The government will look after me if disaster strikes."

Expanding the Safety Net

This entitlement mentality ensured that each new tragedy would generate fresh demands to expand the safety net. In Katrina's aftermath, those demands centered on State Farm [insurance company], which dared to deny certain claims under homeowners policies that covered wind damage but expressly excluded floods. Mississippi's attorney general immediately sued to void flood exclusion clauses as "unconscionable" and "contrary to public policy" and even launched a criminal investigation of State Farm's claims adjusting practices.

[In 2007], a jury inflamed by adverse public opinion awarded $1 million in punitive damages against State Farm for having stood on its contract rights in a dispute involving a single house. That case was recently reversed on appeal, but the victory is cold comfort for State Farm, which in the meantime elected prudently to calm the litigation storm by paying tens of millions of dollars to settle claims for unproven wind damage. Voila! The safety net had a brand new strand, woven at the insurance company's expense.

Disgusted, State Farm announced [later] that it would cease writing new homeowners policies in Mississippi.

The solution is to replace the prevailing entitlement mentality with a free market in disaster prevention, insurance, and recovery.

As more private insurers withdraw from high-hazard areas—or raise their rates to reflect the staggering legal and public relations costs of offering disaster insurance—a predictable lament arises: the free market has failed, and government must fill the vacuum so that the statist safety net remains strong. Thus it surprises no one to hear Florida [governor] Charlie Crist challenging presidential candidates to support creation of a federal catastrophic fund that would keep insurance premiums artificially low in disaster-prone areas across the country.

A Free-Market Solution

But the solution is not more of the market distortions and perverse incentives that have lured so many people into harm's way. The solution is to replace the prevailing entitlement mentality with a free market in disaster prevention, insurance, and recovery.

In a free market—without tax-paid levees, government disaster relief, or subsidized insurance—anyone who contemplates building or buying property in a high-hazard area will need to face hard facts about the local history of natural disasters, the efficacy and cost of preventive measures, and the availability of insurance.

For example, the high price—or total unavailability—of private insurance will resound like a clanging alarm bell, signaling the market's objective view that a particular building plan is abnormally risky compared to less dangerous locales.

With their own lives and wealth at stake, people will have every incentive to evaluate risks objectively. And if hardy souls still choose to occupy and fortify New Orleans, or build on an earthquake fault, or live in a tornado alley, the risk and reward will be theirs alone. No longer will government make disasters more disastrous by pretending that citizens have a right to defy the forces of nature at others' expense.

Organizations to Contact

The editors have compiled the following list of organizations concerned with the issues debated in this book. The descriptions are derived from materials provided by the organizations. All have publications or information available for interested readers. The list was compiled on the date of publication of the present volume; names, addresses, and phone numbers may change. Be aware that many organizations take several weeks or longer to respond to inquiries, so allow as much time as possible.

American Red Cross
2025 E Street NW, Washington, DC   20006
(202) 303-5000
Web site: www.redcross.org

The American Red Cross is not a government agency, but it was chartered by Congress in 1905 to "carry on a system of national and international relief in time of peace and apply the same in mitigating the sufferings caused by pestilence, famine, fire, floods, and other great national calamities, and to devise and carry on measures for preventing the same." In times of disaster, the Red Cross focuses on meeting people's immediate disaster-caused needs by providing shelter, food, and physical and mental health services. The Red Cross Web site provides information for individuals and families about preparing for disasters, coping with disasters, and maintaining safety during the disaster recovery process.

Disaster Preparedness and Emergency Response Association (DERA)
PO Box 797, Longmont, CO   80502
e-mail: dera@disasters.org
Web site: www.disasters.org

The Disaster Preparedness and Emergency Response Association is a nonprofit, nongovernmental organization founded in

1962 to link professionals, volunteers, and organizations active in all phases of disaster preparedness and emergency management. DERA provides both professional support and disaster service. Its Web site contains a library of materials, including back issues of the group's monthly newletter *DisasterCom*.

Federal Emergency Management Agency (FEMA)
500 C Street SW, Washington, DC 20472
(800) 621-3362 • fax: (800) 827-8112
Web site: www.fema.gov

FEMA, part of the U.S. Department of Homeland Security, is the national agency responsible for responding to and reducing loss of life and property from all hazards, including natural disasters, acts of terrorism, and other human-caused disasters. FEMA often works in partnership with state and local emergency management agencies, other federal agencies, and the American Red Cross. The FEMA Web site offers information about different types of disasters, the disaster response and recovery process, and tips for individuals and families coping with disasters.

National Oceanic and Atmospheric Administration (NOAA)
1401 Constitution Ave. NW, Room 5128
Washington, DC 20230
(202) 482-6090 • fax: (202) 482-3154
e-mail: outreach@noaa.gov
Web site: www.noaa.gov

NOAA is a federal agency that provides citizens, emergency managers and planners, and decision-makers with reliable information about changes in the environment, including severe storms, hurricanes, tornadoes, fires, extreme heat, tsunamis, floods, and solar flares. NOAA's National Weather Service is the official voice of the United States government for issuing warnings during life-threatening weather situations. NOAA's Web site provides detailed reports on the environment, weather, and natural disasters.

Noah's Wish
PO Box 4288, El Dorado Hills, CA 95762
(916) 939-9474 • fax: (916) 939-9479
e-mail: info@noahswish.info
Web site: www.noahswish.org

Noah's Wish is a nonprofit organization that helps animals during disasters and tries to mitigate the impact of disasters on animals through educational outreach programs. The group's Web site contains photo galleries of a number of disasters and provides general information about disaster preparedness, planning, and response from the perspective of animal safety and rescue.

Oxfam International
226 Causeway Street, 5th Floor, Boston, MA 02114-2206
(617) 482-1211 • fax: (617) 728-2594
e-mail: info@oxfamamerica.org
Web site: www.oxfam.org

Oxfam International is a consortium of organizations that work together around the world to bring justice and better living conditions and opportunities to the poor. Part of this work is keeping people safe and helping them recover from natural disasters. The group's Web site includes press releases and news stories about disasters and disaster response.

ReliefWeb
United Nations Office for the Coordination
    of Humanitarian Affairs
New York, NY 10017
(212) 963-1234
Web site: www.reliefweb.int

ReliefWeb is a Web site administered by the United Nations Office for the Coordination of Humanitarian Affairs to provide information on humanitarian emergencies and disasters. The site is designed specifically to assist the international humanitarian community in effective delivery of emergency assistance, and it provides timely, reliable and relevant information as events unfold. Users can find updates on the latest natural disasters and other emergencies and can access reports and articles about disaster policy and issues. Examples of recent reports include "Damage, Needs or Rights? Defining What Is Required After Disaster," and "Defusing Disaster, Reducing the Risk: Calamity Is Unnatural."

Bibliography

Books

G. Elaine Acker and Pets America — *Pet First Aid and Disaster Response Guide: Critical Lessons from Veterinarians*. Sudbury, MA: Jones and Bartlett, 2008.

Samia Amin and Markus Goldstein — *Data Against Natural Disasters: Establishing Effective Systems for Relief, Recovery, and Reconstruction*. Washington, DC: World Bank, 2008.

Philip E. Auerswald et al. — *Seeds of Disaster, Roots of Response: How Private Action Can Reduce Public Vulnerability*. New York: Cambridge University Press, 2008.

Lee Clarke — *Worst Cases: Terror and Catastrophe in the Popular Imagination*. Chicago: University Of Chicago Press, 2005.

Stephen E. Flynn — *The Edge of Disaster: Rebuilding a Resilient Nation*. New York: Random House, 2007.

Mohamed Gad-el-Hak — *Large-Scale Disasters: Prediction, Control, and Mitigation*. New York: Cambridge University Press, 2008.

Chester Hartman and Gregory D. Squires — *There Is No Such Thing as a Natural Disaster: Race, Class, and Hurricane Katrina*. New York: Routledge, 2006.

R.P. Maiden et al.   *Workplace Disaster Preparedness, Response, and Management.* New York: Routledge, 2007.

David McEntire   *Disaster Response and Recovery: Strategies and Tactics for Resilience.* New York: Wiley, 2006.

James F. Miskel   *Disaster Response and Homeland Security: What Works, What Doesn't.* Westport, CT: Praeger Security International, 2008.

Roger del Moral and Lawrence R. Walker   *Environmental Disasters, Natural Recovery and Human Responses.* New York: Cambridge University Press, 2007.

Charles Perrow   *The Next Catastrophe: Reducing Our Vulnerabilities to Natural, Industrial, and Deliberate Disasters.* Princeton, NJ: Princeton University Press, 2007.

Amanda Ripley   *The Unthinkable: Who Survives When Disaster Strikes—and Why.* New York: Crown, 2008.

Michael E. Whitman and Herbert J. Mattord   *Principles of Incident Response and Disaster Recovery.* Boston: Course Technology, 2006.

Periodicals

Clifton Barnes   "Disaster Response: Lessons—and an Update—from Hurricane Katrina," *Bar Leader*, September–October 2008.

Paige Bowers   "Organizing Disaster," *Time*, February 29, 2008.

Joe W. Craver   "Six Months After Fires, It's Time to Better Prepare," *San Diego Business Journal*, May 12, 2008.

Shana R. Deitch   "Congress, GAO Concerned About Nation's Disaster Response Capability," *Contingency Planning & Management*, August 7, 2007.

Libby George   "When Disaster Strikes and the Response Fails," *CQ Weekly*, July 7, 2008.

Mathias Lemmens   "Disaster Response Starts with a Map," *GIM International*, April 2007.

Karen Mathews   "Disaster Survey Finds US Ill Prepared," *Time*, September 12, 2008.

John McQuaid   "Storm Warning: The Unlearned Lessons of Katrina," *Mother Jones*, August 26, 2007.

New York Times   "Faking the Katrina Inquiry," September 26, 2005.

Neal Peirce   "Tackling the Challenges of Fire and Water on a Warming Globe," *Government Finance Review*, December 1, 2007.

Catharine Skipp   "The Holdouts: What Motivates Some People to Ignore Evacuation Orders and Warnings of Hurricanes and Other Disasters?" *Newsweek*, September 12, 2008.

Bill Terry — "Wildland Fire Management: Stay or Go When Fire Threatens?" *NFPA Journal*, January–February 2005.

Breanne Wagner — "Damage Control: FEMA on a Mission to Regain Credibility," *National Defense*, July 2008.

Internet Sources

ABC News — "What Can Wal-Mart Teach FEMA About Disaster Response? Many Major Companies Provided Disaster Relief Faster than Government Agencies," September 29, 2005. http://abcnews.go.com/WNT/HurricaneRita/Story?id=1171087&page=1.

Thomas W. Brunner and Dale E. Hausman — "Natural Disasters Pose Difficult Challenges for Insurers," Wiley Rein LLP, September 8, 2005. www.wileyrein.com/publication.cfm?publication_id=12319.

CNN.com — "Report: Major Disaster Would 'Overwhelm' Aid Groups," September 18, 2008. www.cnn.com/2008/US/09/18/disaster.aid/.

Naomi Klein — "Pay to Be Saved: The Future of Disaster Response," Common Dreams, August 28, 2006. www.commondreams.org/views06/0829-23.htm.

Cheryl Pellerin   "Disaster Management Moving from Response to Prevention," America.gov, June 10, 2005. www.america.gov/st/ washfile-english/2005/June/ 20050610135814nirellep0. 1706812.html.

Science Daily   "Disasters Getting Worse—US Government Must Be Better Prepared, Report Urges," November 12, 2007. www.sciencedaily.com/ releases/2007/11/071112073337.htm.

Science Daily   "Seismic Network Could Improve Disaster Response," February 12, 2005. www.sciencedaily.com/ releases/2005/02/050211094621. htm.

Index